FOOD SCIENCE AND TECHNOLOGY

GLUTEN:
PROPERTIES, MODIFICATIONS
AND DIETARY INTOLERANCE

FOOD SCIENCE AND TECHNOLOGY

Additional books in this series can be found on Nova's website under the Series tab.

Additional E-books in this series can be found on Nova's website under the E-books tab.

FOOD AND BEVERAGE CONSUMPTION AND HEALTH

Additional books in this series can be found on Nova's website under the Series tab.

Additional E-books in this series can be found on Nova's website under the E-books tab.

FOOD SCIENCE AND TECHNOLOGY

GLUTEN: PROPERTIES, MODIFICATIONS AND DIETARY INTOLERANCE

DIANE S. FELLSTONE
EDITOR

Nova Science Publishers, Inc.
New York

Copyright © 2011 by Nova Science Publishers, Inc.

All rights reserved. No part of this book may be reproduced, stored in a retrieval system or transmitted in any form or by any means: electronic, electrostatic, magnetic, tape, mechanical photocopying, recording or otherwise without the written permission of the Publisher.

For permission to use material from this book please contact us:
Telephone 631-231-7269; Fax 631-231-8175
Web Site: http://www.novapublishers.com

NOTICE TO THE READER

The Publisher has taken reasonable care in the preparation of this book, but makes no expressed or implied warranty of any kind and assumes no responsibility for any errors or omissions. No liability is assumed for incidental or consequential damages in connection with or arising out of information contained in this book. The Publisher shall not be liable for any special, consequential, or exemplary damages resulting, in whole or in part, from the readers' use of, or reliance upon, this material. Any parts of this book based on government reports are so indicated and copyright is claimed for those parts to the extent applicable to compilations of such works.

Independent verification should be sought for any data, advice or recommendations contained in this book. In addition, no responsibility is assumed by the publisher for any injury and/or damage to persons or property arising from any methods, products, instructions, ideas or otherwise contained in this publication.

This publication is designed to provide accurate and authoritative information with regard to the subject matter covered herein. It is sold with the clear understanding that the Publisher is not engaged in rendering legal or any other professional services. If legal or any other expert assistance is required, the services of a competent person should be sought. FROM A DECLARATION OF PARTICIPANTS JOINTLY ADOPTED BY A COMMITTEE OF THE AMERICAN BAR ASSOCIATION AND A COMMITTEE OF PUBLISHERS.

Additional color graphics may be available in the e-book version of this book.

LIBRARY OF CONGRESS CATALOGING-IN-PUBLICATION DATA

Gluten : properties, modifications and dietary intolerance / Editor: Diane
S. Fellstone.
p. cm.
Includes index.
ISBN 978-1-61209-317-8 (hardcover)
1. Gluten. 2. Gluten-free foods. I. Fellstone, Diane S.
QK898.G49G575 2011
613.2'82--dc22
 2010048370

Published by Nova Science Publishers, Inc. † New York

Contents

Preface		vii
Chapter 1	Gluten-Free Bread: Economic, Nutritional and Technological Aspects *Maria Teresa Pedrosa Silva Clerici and Yoon Kil Chang*	1
Chapter 2	Impact of Nitrogen and Sulfur Fertilization on Gluten, Composition, and Baking Quality of Wheat *Dorothee Steinfurth, Karl H. Mühling and Christian Zörb*	27
Chapter 3	Functional Gluten Alternatives *Aleksandra Torbica, Miroslav Hadnađev, Tamara Dapčević Hadnađev and Petar Dokić*	49
Chapter 4	Chiaroscuro of Standardization for Gluten-Free Foods Labeling and Gluten Quantitation Methods *A. M. Calderón de la Barca, E. J. Esquer-Munguía and F. Cabrera-Chávez*	75
Chapter 5	Effect of Heat on Gluten *Costas E. Stathopoulos and Quan V. Vuong*	89
Chapter 6	Modern Concepts Pathogenesis of Celiac Disease: From Gluten to Autoimmunity *Asma Ouakaa-Kchaou*	101
Chapter 7	Emulsion Properties of Different Protein Fractions from Hydrolyzed Wheat Gluten *S. R. Drago, R. J. González and M. C. Añón*	113
Chapter 8	Techno-functional Properties from Hydrolyzed Wheat Gluten Fractions *S. R. Drago, R. J. González and M. C. Añón*	133

Chapter 9	Gluten-Free Diet in Children and Adolescents with Celiac Disease *Gian Vincenzo Zuccotti, Dario Dilillo, Fabio Meneghin and Cecilia Mantegazza*	**161**
Index		**183**

PREFACE

This new book presents current research in the study of gluten, including the economic, nutritional and technological aspects of gluten-free bread; impact of nitrogen and sulfur fertilization on gluten composition; functional gluten alternatives; effect of heat on gluten and the gluten-free diet in children with celiac disease.

Chapter 1 – This chapter presents a detailed description of previously published works of the economical, nutritional and technological aspects of the research and production of gluten-free bread, which can benefit not only countries that do not produce wheat, but also celiac patients, whose life-long diet completely excludes foods containing wheat, oat, barley, rye and triticale. The real quality of the bread and wheat currently consumed will be investigated in this chapter: the years of research, massive financing by the wheat-exporting countries and private enterprises, high volume of production and intense marketing, all of which contributed to the rise in the global consumption of bread. The quality of gluten-free bread is not yet like that of wheat bread, but this research targets the improvement of the three-dimensional polysaccharide or protein network structure through physical and chemical modifications of starch or protein, so that they may result in wheat-free products of significant technological quality.

Chapter 2 – Wheat has a broad genetic potential for the expression of various gluten proteins and is strongly influenced by the environment, e. g. with respect to the fertilization of nitrogen (N) and sulfur (S) during wheat plant development. Nitrogen and S availability markedly change wheat flour and baking quality. However, without adequate management of N and S fertilization, the genetic potential of wheat cannot be exploited. Nitrogen fertilization primarily affects the concentration of gliadines and glutenines as it is source-regulated. Storage protein concentration, particularly of gliadine and glutenine, is highly correlated with loaf volume. The use of three instead of one dressing of fertilizer, especially as a late dressing at ear emergence, the gliadine and glutenine concentration of the grain and, furthermore, loaf volume can be enhanced. However, not only quantity, but also the quality of gluten proteins is important for their composition and characteristics. Despite protein amount being only slightly influenced by increasing S fertilization levels, the composition of gluten proteins is highly affected. Increasing S availability for plants enhances S-containing amino acids, namely cysteine and methionine, which are foremost in S-rich gluten proteins. For an improvement of baking quality, a high concentration of S-rich proteins is of particular importance in order to form an appropriate gluten network within the dough. Additionally, a late S dressing can further improve baking quality, since such a dressing increases the synthesis of high molecular weight (HMW) glutenins that considerably affect baking quality in wheat.

S-rich components such as sulfate and glutathione are, moreover, important molecules for the transport and supply of S to the grain. Both these molecules represent important S transport forms from source to sink organs. The potential to synthesize large amounts of storage protein depends on the ability of the plant to take up and to transport S into sink organs such as ears and kernels. Sulfur and glutathione are therefore important molecules functioning as actuators during plant growth and grain development. Furthermore, glutathione can function as internal plant signal for the S status of the plant. In terms of baking quality, glutathione is able to interact with gluten proteins responding in a change of existing disulfide bonds between gluten proteins and resulting in a different rheological property and baking quality. Concerning the interaction of N and S in wheat, a well-balanced N/S ratio is exceedingly important for suitable gluten and baking quality. If high N fertilization is accompanied by inadequate S fertilization, S deficiency is provoked, resulting in changed gluten protein composition and a loss of nutritional quality of the grains.

Chapter 3 – The demand for gluten-free products is ascending steadily, paralleling the increase in prevalence and incidence of celiac disease and other allergic reactions or intolerances to gluten consumption. The replacement of gluten presents a major technological challenge, as its visco-elastic properties largely determine the breadmaking performance of wheat flour. Various gluten-free formulations apply mixtures of rice or corn flour and different hydrocolloids or starches to mimic the unique properties of gluten. However, gluten-free products containing gums and starches as gluten replacements lack in essential nutrients. Therefore, replacing standard gluten-free formulations with those made from alternative grains like buckwheat may impart nutritional benefits. This chapter reviews the literature on gluten-free bakery products, including both hydrocolloid-based and pseudo-cereal-based formulations. A special emphasis is given to ongoing research in the authors laboratory related to gluten-free bread and cookies containing rice and buckwheat flour. The optimal gluten-free formulations were created by comparing the rheological properties of different rice and buckwheat flour mixtures to properties of wheat flour assessed by using Mixolab. Subsequently, rice, buckwheat and wheat flours were evaluated by electrophoretic and electron-microscopic analysis. Moreover, the influence of buckwheat flour type and content on rheological, textural, sensory properties of gluten-free dough, bread and cookies was investigated.

Chapter 4 – Celiac disease (CD) is an autoimmune enteropathy characterized by intolerance to wheat gluten. CD patients must adhere to a strict lifelong gluten-free diet, excluding all food products containing wheat and taxonomically related cereals. The CD prevalence is 1% in any population over the world and apparently it is increasing. Thus, the gluten content must be regulated in specialties for gluten intolerant patients by reliable and sensitive methods. To reach this objective, several procedures have been used including immunological and non-immunological methods. Additionally to limitations on gluten quantitation due to the principles of the tests, there are limitations due to the methods of proteins' extraction. In this chapter, troubles and trends for gluten quantitation in gluten-free foodstuffs are discussed. Also differences and agreements on the regulations of gluten-free labeling as well as the related basic definitions are shown. Finally, it is stated the consideration of the industry needs, current scientist knowledge and the safety of gluten intolerant patients for a good regulation and international trade of gluten-free foodstuffs.

Chapter 5 – The effect of processing, and heat in particular, on the wheat gluten proteins can be difficult to explain due to its complex, and often unusual, rheological and biochemical properties. Heat denaturation of wheat gluten proteins and the accompanying rheological changes, together with a number of interactions, such as hydrogen bonding, SS bonding and hydrophobic interactions have an effect on the native structure of the protein. During the heat treatment of gluten, denaturation, aggregation and cross-linking all combine to give rise to a series of changes that affect rheological and biochemical properties alike. Different components of gluten might exhibit different responses to heat treatment, based on their parent wheat variety, their size, their composition, or the environment the heat treatment took place. Today's food scientists are yet to fully understand all the interactions and mechanisms involved in the effect of heat on gluten but this field of research has grown enormously over the last few decades and continuously expands offering us a better insight and understanding.

Chapter 6 – Background: Celiac disease, also known as gluten-sensitive enteropathy and non-tropical sprue, is a prevalent autoimmune disorder that is triggered by the ingestion of gluten and related prolamins in genetically susceptible individuals. The classic celiac lesion occurs in the proximal small intestine with histological changes of intestinal villous atrophy, crypt hyperplasia, intraepithelial lymphocytosis and leukocyte infiltration of the lamina propria. The pathogenic mechanisms in this disease are not yet well understood, but it is clear that genetic, environmental and immunological factors play a role.

Aim: To provide an evidence-based overview of the pathogenesis of celiac disease.

Methods: Review based on relevant medical literature.

Results: Celiac disease is uniquely characterized by a defined trigger (gluten proteins from wheat and related cereals), the necessary presence of HLA-DQ2 or HLA-DQ8, and the generation of circulating autoantibodies. Celiac disease has become one of the best-understood HLA-linked disorders. Well-identified haplotypes in the human leukocyte antigen (HLA) class II region (either DQ2 [DQA*0501-DQB*0201] or DQ8 [DQA*0301-DQB1*0302]) confer a large part of the genetic susceptibility to celiac disease. The immune response in celiac disease involves the adaptive, as well as the innate, and is characterized by the presence of anti-gliadin and anti-transglutaminase antibodies, lymphocytic infiltration in the epithelial membrane and the lamina propria, and expression of multiple cytokines and other signaling proteins.

Conclusion: Gluten-free diet is currently the only effective mode of treatment for celiac disease; nevertheless, there is a growing demand for alternative treatment options. Better understanding of the mechanism of the disease is likely to add other choices for therapy in the future.

Chapter 7 – Many protein sources that are found in the market are obtained as by-products and there is a great interest in using them as protein ingredients with adequate functionality for food formulation. Structure modification allows to add value and to diversify their uses. The viscoelastic properties of wheat gluten have restricted its use in baked products, and the diversification of gluten applications depends of the improvement of its solubility in a wider pH range. An alternative for that is the enzymic hydrolysis.

The objective of this work was to evaluate emulsion properties of protein fractions obtained by extracting at 3 pH different hydrolyzed gluten samples.

Hydrolyzates were made using two commercial enzymes (acid and alkaline proteases) to reach 3 different hydrolysis degrees (DH) for each enzyme. Extracts were obtained at 3 different pHs (4, 6.5 and 9) and were diluted to a protein concentration of 4 g/l. Each extract was used to make the corn oil: extract emulsions (25:75, W/W).

Emulsion capacity was determined by measuring droplet size distribution and the stability using a vertical scan macroscopic analyzer.

Regarding emulsion capacity, multifactor ANOVA (factors: pH and DH) made for droplet size distribution parameters showed that there were no differences between samples.

Regarding stability evaluation, alkaline protease extract emulsions were more stable in particle migration phenomena by creaming, but showed higher coalescence rates than those corresponding to acid protease extract emulsions.

It was also observed that for both enzymes, as DH increases, coalescence rates decrease for the 3 pHs extracts and creaming rates increase for pH 4 and 9 extracts. In the case of pH 6.5, acid protease extracts emulsions showed a clear creaming instability by flocculation, probably due to the electrical charges suppression of the peptides adsorbed at the interface, since 6.5 is a pH near the isoelectric point of gluten proteins. It is suggested that acid protease extract emulsions showed a certain degree of bridging flocculation. This caused higher creaming rates but a lower coalescence as a consequence of the bridging. The authors conclude that although pH of the extraction and DH did not affect emulsion capacity, emulsion stability depended on the pH, DH and the enzyme used.

Chapter 8 – Many protein sources that are found in the market are obtained as by-products and there is a great interest in using them as protein ingredients with adequate functionality for food formulation. Structure modification permits one to add value and to diversify their uses. The diversification of wheat gluten applications depends on the improvement of its solubility in a wider pH range. One of the alternatives that allow protein modification in these products is the enzymic hydrolysis. The objective of this work was to evaluate foaming properties of protein fractions obtained by extracting, at 3 pH, different hydrolyzed gluten samples. Two commercial enzymes (acid -Ac- and alkaline -Al- proteases) were used to reach 3 different hydrolysis degrees (DH). Extracts pHs (4, 6.5 and 9) were diluted to a protein concentration of 4 g/l. RP-HPLC, free amino groups content, sulphydryl and disulfur content, average peptide chain length were used to characterize each extract. Foam was produced by sparging nitrogen at a known rate through a dilute protein solution. The maximal volume of liquid incorporated into the foam (Vmax) and the rate of liquid incorporation into the foam (Ri) were determined and used as indicators of foaming capacity. The times for half-drainage of the liquid that was incorporated into the foam at the end of the bubbling period (t1/2) and the rate of liquid drainage from the foam were also measured. Regarding Ri, all pH 4 extracts from hydrolyzed samples showed higher Ri than an un-hydrolyzed sample. Extracts from Al hydrolyzed extracts showed higher Ri than those from Ac. In the case of Al extracts, an inverse relation between DH and Ri was observed, but practically no influence of DH on Ri, was observed in the case of Ac extracts. For pH 6.5 extracts, the relation between DH and Ri were in opposite directions, depending on the enzyme, for Ac, Ri decreased with DH, while for Al, Ri increased with DH. At this pH, it was observed that the extracts which foamed more quickly, were those with the highest times for half-drainage of the liquid (t1/2). Some foam parameters correlated between themselves, depending on the extracts. Foaming capacity and stability depend on pH, DH and enzyme and it was possible to correlate parameters with composition evaluated by RP-HPLC.

Chapter 9 – Celiac disease (CD) is defined as a permanent sensitivity to the gluten in wheat and related proteins found in barley and rye, which occurs in genetically susceptible individuals and affects 0.5-1% of the general population worldwide. Currently, the only available treatment is lifelong adherence to a gluten-free diet (GFD). There is evidence that untreated CD is associated with a significant increase in morbidity and mortality. It has been demonstrated that even small amounts of ingested gluten can lead to mucosal changes upon intestinal biopsy. Previously, products containing less than 200 ppm (<200 mg/kg) were regarded as gluten-free. Currently, less than 20 ppm (<20 mg/kg) is being considered in the proposed Codex Alimentarius Guidelines to define gluten-free. In the USA, the national food authority has recently redefined their definition of "gluten-free" with a threshold of no gluten. The use of oats is not widely recommended because of concerns about potential contamination during the harvesting and milling process, so unless the purity of oats can be guaranteed, its safety remains questionable.

A GFD has both lifestyle and financial implications for patients; thus, it can adversely impact their quality of life, such as difficulties eating out, a negative impact on career and family life, anxieties about social difficulties, and feeling different. Though the compliance to a GFD started in childhood is very high, the percentage of adolescents with CD who strictly follow a GFD varies from 43% to 81%; a greater adherence is found within patients with typical symptoms. Moreover, the alimentary habits of healthy adolescents exhibit nutritional imbalances with a high consumption of lipids and proteins and a low consumption of carbohydrates, calcium, fiber and iron; several studies show that adherence to a strict GFD worsens the nutritional imbalances of an adolescent with CD.

Gluten-free products are often low in B and D vitamins, calcium, iron, zinc, magnesium, folate, thiamine, riboflavin, and niacin. Very few commercially available gluten-free products are enriched. Vitamin and mineral supplementation can be useful adjunct therapy for a GFD. Patients inadequately treated have low bone mineral density, imbalanced macronutrients, low fiber intake, and micronutrient deficiencies.

Moreover, recent research has found that adults on a strict GFD for years have high total plasma homocysteine levels; this finding is associated with a high prevalence of being overweight and represents an increased risk for metabolic and cardiovascular disease.

In: Gluten: Properties, Modifications and Dietary Intolerance
Editor: Diane S. Fellstone
ISBN: 978-1-61209-317-8
©2011 Nova Science Publishers, Inc.

Chapter 1

GLUTEN-FREE BREAD: ECONOMIC, NUTRITIONAL AND TECHNOLOGICAL ASPECTS

Maria Teresa Pedrosa Silva Clerici[*1] *and Yoon Kil Chang*[‡2]

[1]Federal University of Alfenas (UNIFAL-MG), Brazil, Alfenas, MG, Brazil
[2] State University of Campinas (UNICAMP), Department of Food Technology – Food Engineering Faculty, Campinas, SP, Brazil

ABSTRACT

This chapter presents a detailed description of previously published works of the economical, nutritional and technological aspects of the research and production of gluten-free bread, which can benefit not only countries that do not produce wheat, but also celiac patients, whose life-long diet completely excludes foods containing wheat, oat, barley, rye and triticale. The real quality of the bread and wheat currently consumed will be investigated in this chapter: the years of research, massive financing by the wheat-exporting countries and private enterprises, high volume of production and intense marketing, all of which contributed to the rise in the global consumption of bread. The quality of gluten-free bread is not yet like that of wheat bread, but this research targets the improvement of the three-dimensional polysaccharide or protein network structure through physical and chemical modifications of starch or protein, so that they may result in wheat-free products of significant technological quality.

1. INTRODUCTION

Due to its color, flavor and aroma, bread has become one of the world's most consumed foods. Wheat, considered a premium variety of grain, currently stands as the only one of its

---

*Federal University of Alfenas (UNIFAL-MG), Brazil, Rua Gabriel Monteiro da Silva 700, Centro, CEP 37130-000, Alfenas, MG, Brazil, E-mail: maria.pedrosa@unifal-mg.edu.br, mtcleric@fea.unicamp.br, Telephone: +55-35-3299-1110

‡ State University of Campinas (UNICAMP), Department of Food Technology – Food Engineering Faculty, P.O. Box 6121, Zip Code 13083-862, Campinas, SP, Brazil, E-mail: yokic@fea.unicamp.br, Telephone: +55-19-3521-4001

kind to possess a net of gluten-forming proteins with enough elasticity and extensibility to retain gas and expand itself in products such as bread, which attributes a generally pleasant form and appearance to the final product. This unique trait resulted ultimately in large investments in the production of wheat in order to provide for global demand [26, 37, 98, 127].

In the late 1960's and early 1970's, countries that produced wheat with good properties for bread manufacturing used an economic strategy in which they exported both their wheat and the low-investment technology required for manufacturing wheat-based products (breads, cakes, pasta, cookies); in doing so, they promoted the consumption of these final products [38, 39, 40, 41].

Consequently, the need to import wheat grew expressively for non-wheat-producing countries, resulting in a decline in the consumption of products made with native grains, and therefore in the local farming of these very grains [47].

Reinstating the habit of consuming native tubers and grains (such as cassava, corn, and rice), and developing products with total or partial wheat-substitutes became increasingly difficult due to the high availability and technological quality of wheat-based products [31, 98].

The change in eating habits, along with the increase in bread consumption in several daily meals, made bread an important source of energy for a large part of the world's population. However, with scientific progress made in the diagnosis of various illnesses, it was observed that many people suffered from wheat allergy and celiac disease, which is genetic. For both of these diseases, treatment is a life-long diet completely free of wheat-based products.

Since the supply of gluten-free products in the market is small, celiac and wheat allergy patients suffer greatly from a social to a nutritional level due to their conditio. Not only is there a lack of investment in the research and development of new products, but the gluten-free products currently available in the market have low-quality texture, color and flavor and are expensive as well. These factors limit their consumption among celiac patients.

To reinforce this, the Brazilian Celiac Association (ACELBRA, 2) published online a research study that indicated what kind of food the celiac patients registered to their website would most like to readily find. The results were: bread (47%), pasta (21%), cookies and crackers (21%) and pizza (11%).

Through the study of the economical, nutritional and technological aspects of previous research and production of gluten-free bread, this chapter intends to contribute to this area of research, which shows a promising and challenging future for researchers who seek a substitute for wheat gluten.

2. THE IMPORTANCE OF BREAD IN THE DIET

The versatility in bread production, combined with its softness, the fact that it's easy to chew, and the number of possible sweet and salty fillings for it, makes bread a product consumed in various occasions: from a simple sandwich to party appetizers.

Nutritionally, bread has become a great source of energy, especially in carbohydrates, for a large part of the world's population. This is such that the World Health Organization

(WHO) recommends that each person consume 60kg of bread per year and the Food Agricultural Organization (FAO) recommends 50kg per year [46].

As one of the products that continuously receive large technological investments, bread has several manufacturing processes. Although indirect and direct processes can produce, through small formulation modifications, a wide variety of kinds of bread, the continuous process is limited to white bread and is used for the manufacturing of large volumes. If small adjustments can be made, usually in the formulation, all three processes can result in final products with different sensory characteristics (format, texture, volume, color, etc). The larger portion of consumption includes French bread, white bread, hamburger and hotdog rolls, and buns [127].

Since the manufacturing of bread of good technological and sensorial quality is directly linked to the gluten-forming proteins in wheat, many investments have been made to predict and determine the "strength" of wheat flour.

The expression "strength of flour" is normally used to designate a certain flour's greater or lesser capacity to suffer mechanical treatment while being mixed with water, associated with the gluten-forming protein's greater or lesser capacity to absorb water, and combined with the capacity to retain carbonic gas; all resulting in a good final bakery product. Wheat flour for bread should have: protein content between 11 and 14%, high water-absorption capacity, gluten with medium levels of elasticity and extensibility, and high resistance to mechanical treatment. Suitable flour is capable of producing bread with good volume, texture, color and flavor [17, 26, 37, 102, 128].

Due to wheat flour's unique characteristics, the milling industry and researchers at universities and institutes have been developing methods to evaluate its quality [1, 35, 92]. Today, the most used methods for the evaluation of grains and flour include: test weight; thousand kernel weight; hardness index in single kernel characterization system (SKCS); falling number grain; experimental milling or flour yield in Brabender Quadrumat Senior mill; flour moisture content [1]; gluten analysis: wet gluten, dry gluten and, gluten index [1]; alveography and farinography methods for flour [1], and the baking test [35] and cookie test [1].

The final price of the wheat and its flour depends on the strength of its gluten-forming proteins. When certain wheat flour does not present the necessary quality for bread manufacture, many additives, enzymes and ingredients are used to improve the gluten's strength, for example:

- Additives: oxidants such as benzoyl peroxide [32], ascorbic acid [57]
- Enzymes: glucose oxidase [14, 124, 100,101], transglutaminase [7, 119]
- Ingredients: vital gluten, a kind of gluten extracted from wheat by use of processes that do not affect its technological properties [121]

Many other methods, ingredients and additives are being researched and are undergoing industrial development, which demonstrates the force that the wheat market has in directing scientific research.

Due to the fact that bread is also well known for its high glycemic index and low fiber content [53, 76, 108], many investments are being made in the research and development of breads with health benefits, such as the use of whole grains and/or fibers, the substitution of sugar and/or fats, the addition of other whole grains, and the addition of prebiotics, among other components that improve the nutritional value of bread. Many of these products are commercially available in a great number of countries.

The addition of a single ingredient is thoroughly researched, because when the dilution of gluten-forming proteins occurs, the negative impact on the final product's quality can halt the consumer's acceptance of the product [38, 39, 40, 41, 57, 61].

The formation of gluten, vital to the quality of bread, has always been under study. The theories that try to explain this phenomenon are summarized in item 3 of this chapter.

3. THEORIES REGARDING THE FORMATION OF GLUTEN

The attempt to understand the formation of gluten is long-standing and many studies that have looked into gluten-forming proteins have taken into consideration their solubility and extraction. One of the most significant studies was that of Nasmith [81], which not only discussed information given by various authors and carried out experiments, it also concluded that gluten formation is not merely a mechanical mixture of gliadin with glutenin; a definite physical state of the two mixing substances is necessary, especially since coagulated glutenin with gliadin does not form gluten.

Wheat proteins are constituted of soluble or non-gluten-forming proteins and insoluble or gluten-forming proteins. Soluble proteins consist of albumin and globulins and represent approximately 20% of the total proteins. Insoluble proteins are responsible for gluten formation; they consist of gliadin and glutenin, and represent 80% of total proteins [98].

After hydration and mechanical work, the wheat flour forms dough characterized by a high extensibility and low elasticity. In other words, viscoelastic dough that is capable of gas retention produced during dough fermentation and bread baking [98, 127].

Various hypotheses for explaining the outstanding properties of proteins which give the dough produced from wheat the unique viscoelastic characteristics essential for the production of bread were developed and are as follows:

3.1. Sulfahydryl-Disulfide Bonds Hypothesis

In 1957, Goldstein [52] suggested that a sulfahydryldissulfide interchange reaction could explain the elastic properties of the gluten network.

Later on, some authors [8, 9, 105, 106] studied the gluten formation based on the fact that sub-units of gluten-forming proteins connect to each other by intra and intermolecular disulfide bonds during the making of the dough, to form a continuous and coherent phase.

Beckwith *et al.* [9] and Beckwith & Wall [8] studied the reduction and reoxidation of disulfide bonds and sulfahydryl or thiol groups in a glutenin and gliadin solution. They showed that glutenin consists of relatively low molecular weight components cross-linked intermoleccullary by disulfide bonds, while gliadin, which does not change substantially in molecular weight upon reduction, consists of a monomeric unit with no intermolecular disulfide connections.

This theory was proved correct when the solvent used was water, but when an organic solvent was used, the gluten mass did not form.

3.2. Sulfahydryl-Disulfide Biosynthesized Concatenation Hypothesis

Greenwood & Ewart [54] stated that glutenin is the only constituent of the wheat flour proteins that shows any significant viscoelastic properties that are already existent in the wheat kernel, but are activated by wetting and mixing during dough formation.

3.3. Secondary Bonding Aggregation Forces Hypothesis

Bernadin & Kasarda [11, 12] and Kasarda *et al.* [68] based their hypothesis in the microfibrillar aggregate formation of some gliadin and some other endosperm proteins. They suggest that the formation of dough upon mixing wheat flour with water was solely due to the presence of secondary bonding forces, which are involved in the aggregation of wheat protein subunits.

Kasarda *et al.* [68] stated that the sulfahydryl groups and disulfide bonds play an important role in determining the viscoelatic characteristics of dough, but they admit that the details of that role are not clearly established.

3.4. Secondary Bonding Aggregation and Disulfide Bridging Hypothesis

Upon examination of various hypotheses of gluten formation, El-Dash [36] proposed a complementary model for wheat gluten, where proteins unite with intermolecular disulfide bonds, providing bridging between layers formed by secondary bonding, thus creating a three dimensional lattice network. This author subsequently proposed that it is possible to create a three-dimensional network of polysaccharides or starch based solely on hydrogen bonding, and proved to be capable of substituting the gluten network as far as the production of pasta and bread.

The physical properties of hydrated wheat proteins are the result of covalent and non-covalent interactions of wheat gluten proteins. The repeated extension, tearing, and compression during mixing or development alter these interactions. According Robertson [109], specific chemical effects include:

- Disulfide bond disruption
- Chain disentanglement and rupture
- Disulfide-sulfhydryl interchange
- Formation of dityrosine cross-links
- Formation of new disulphide cross-links
- Free radical interactions, and especially
- Reorientation leading to enhanced hydrogen bonding

The study of the mechanisms involved in the formation of gluten led to the development of additives and enzymes capable of increasing the gluten strength of wheat flour that showed weak gluten networks, inadequate for the manufacturing of bread. Research showed that despite the fact that the prolamins were present in other grains, only rye has weak mass formation. Redman and Ewart [105,106] and Ewart [43, 44, 45] suggested that speed of the interchange reaction in rye-flour dough is slow and incomplete.

Taking into consideration the increase in the number of products containing gluten on a worldwide level, and the fact that bread has become a high quality product consumed early on by children, the number of wheat allergy reactions and symptoms of celiac disease is also rising. Diagnosis is quicker and more precise, and health representatives are more aware of the effect celiac disease can have on a patient.

Celiac disease is different from wheat allergy, and will be approached in the following topic of this chapter. Allergic reactions are mediated by the antibody IgE, which attacks the fractions of gliadin from the consumed wheat. One example is wheat-dependent, exercise-induced anaphylaxis (WDEIA). Celiac disease does not cause anaphylactic shock and occurs only in genetically susceptible individuals [90].

4. Celiac Disease

The prevalence of celiac disease in the United States seems to be similar to that of many European countries, which is 0.5% to 1.0% [42, 62]. Assuming that even a half of sufferers are detected, we can estimate that 1 million US citizens should adhere to a gluten-free diet. Brazil does not have official statistics yet, but a research study carried out with adult blood donors showed a proportion of 1 celiac patient for every 214 residents of São Paulo city [48].

Celiac disease is an autoimmune disorder occurring in genetically susceptible individuals, triggered by gluten and related prolamins. Well-identified haplotypes in the human leukocyte antigen (HLA) class II region (either DQ2 or DQ8) confer a large part of the genetic susceptibility to celiac disease [66, 67, 90].

The presentations of celiac disease are (i) typical, characterized mostly by gastrointestinal signs and symptoms; (ii) atypical or extra intestinal, where gastrointestinal signs/symptoms are minimal or absent and a number of other manifestations are present; (iii) silent, where the small intestinal mucosa is damaged and celiac disease autoimmunity can be detected by serology, but there are no symptoms; and, finally, (iv) latent, where individuals possess genetic compatibility with celiac disease and may also show positive autoimmune serology, that have a normal mucosa morphology and may or may not be symptomatic [4, 66, 67, 116].

Although neurological manifestations in patients with established celiac disease have been reported since 1966, it was not until 30 years later that, in some individuals, gluten sensitivity was shown to manifest solely with neurological dysfunction. Furthermore, the concept of extra intestinal presentations without enteropathy has only recently become accepted [58].

The diagnosis of celiac disease still rests on the demonstration of changes in the histology of the small intestinal mucosa. The classic celiac lesion occurs in the proximal small intestine with histological changes of villous atrophy, crypt hyperplasia, and increased intraepithelial lymphocytosis. Currently, serological screening tests are utilized primarily to identify those individuals in need of a diagnostic endoscopic biopsy. The serum levels of immunoglobulin (Ig)A anti-tissue transglutaminase (or TG2) are the first choice in screening for celiac disease, displaying the highest levels of sensitivity (up to 98%) and specificity (around 96%). Anti-endomysium antibodies-IgA (EMA), on the other hand, have close to 100% specificity and a sensitivity of greater than 90% [116].

Screening studies have revealed that celiac disease is most common in asymptomatic adults in the United States. Although considerable scientific progress has been made in understanding celiac disease and in preventing or curing its manifestations, a strict gluten-free diet is the only treatment for celiac disease to date [82].

Biagi & Corazza [13], in a broad review, evaluated mortality rates in different forms of celiac disease, such as symptomatic celiac disease, unrecognized celiac disease, dermatitis herpetiformis and refractory celiac disease. This author noticed that the mortality rate for celiac disease seems to be higher in Southern than in Northern Europe and seems to correlate with 'national' gluten consumption. He explained these differences based on the hypothesis that links mortality rates to the amount of gluten consumed not only after but also before the diagnosis of celiac disease.

According to Pietzak [95] and Pietzak et al. [96], the term gluten should be designated as prolamins found in wheat (gliadins), rye (secalins), barley (hordeins), triticale (rye and wheat hybrid), and in oats (avenins), since these grains also trigger celiac disease when consumed, although there is some controversy regarding oats.

The argument favorable to the use of oats is based on the fact that oats show low toxicity, higher digestibility [63, 67, 93] and if the oats are exempt from contamination from other grains such as wheat, they show no effect on the celiac patient [60, 63]. Some authors, such as Pitzak et al [96], believe that the risk of wheat or other grain contamination of oats is too high, and it is not worth allowing the patient to undergo such unnecessary risk in his diet.

Contaminating oats with wheat was confirmed in a study that evaluated the use of oats and the effect of oats on the symptoms and quality of life of 1,000 randomly selected members of the Celiac Society. They verified that 710 patients (73% with celiac disease and 55% with dermatitis herpetiformis) were currently consuming oats. Patients appreciated the taste, the ease of use, and the low costs; 94% believed that oats diversified the gluten-free diet; 15% of celiac disease and 28% of dermatitis herpetiformis patients had stopped eating oats and the most common reasons for avoiding oats were fear of adverse effects or contamination [93].

Howdle [62] criticizes the studies that analyze the inclusion of oats in the diet because according to him, the tests that are being carried out are using too-small doses of this grain and are not long-term studies, which would guarantee the safe use of oats by the celiac patient.

The prolamins present in rice and corn do not activate the occurrence of celiac disease [66, 67].

The exclusion of gluten from the diet is a formidable task for dietitians, as wheat flour is present in a wide range of products including bread, cakes, biscuits, and pastas [97, 98, 124].

In the future, new forms of treatment may include the use of gluten-degrading enzymes to be ingested with meals, the development of alternative, gluten-free grains by genetic modification, the use of substrates regulating intestinal permeability to prevent gluten entry across the epithelium, and, finally, the availability of different forms of immunotherapy [117].

In many countries, the information "contains gluten" [16] on food labels is mandatory and control analyses have been carried out, although results are not always satisfactory, such as those found by Cawthorn [20]. He evaluated twenty-five South African food products for gluten and wheat to ascertain whether these products may pose a risk to celiac and/or wheat allergic individuals. His findings were that rye and barley flours and two oat products were shown to be contaminated with wheat; ten out of seventeen naturally gluten-free products contained gluten, although in thirteen of these the levels were below 20 mg/kg and the labels of four products were found to be misleading in terms of the gluten and/or wheat claims made.

Concern over products for celiac patients was originally documented in 1963 by Willem-Karel Dicke, a Dutch pediatrician and the first clinician to develop the gluten-free diet, and to prove that certain types of flour cause relapses in celiac disease patients [59].

One of the products that present a great challenge is gluten-free bread, since the difficulty is in finding a substitute for gluten that is capable of forming a three-dimensional network. As previously mentioned, studies published by ACELBRA [2] show that bread is the number one product that celiac patients would like to find easily.

5. GLUTEN-FREE BREAD

Studies directed towards the development of wheat-less bakery products were initially targeted at countries that did not produce wheat for bread manufacturing, such that the FAO [47] carried out a broad study focused on promoting the research and development of gluten-free bread made with the native flours in developing countries that were dependent on imported wheat.

In these studies [47], it was suggested that the use of a raw material for the development of a gluten-free product should obey the following criteria: compatibility and availability. Flour that has good compatibility to substitute wheat should be white, free of strange or strong odors and with a protein level similar to that of wheat. Flour with good availability should be competitive in terms of cost and pricing and be inferior to that of wheat.

Finally, the proposal was that research involving the substitution of wheat in bread should obey the local criteria of each country, based ultimately on the choice of grain or tuber, which can be planted in sufficient quantity to supply the development and production of wheat-less bread [47]. In other words, the farming of a crop with a new purpose requires adequate planning: infrastructure for the production, storage, processing and transportation of tons of a new crop [31]. The substitution of wheat for the flour of other grains, roots and tubers depends on the production of each country.

Flour obtained from grains showed advantages such as: high yield during production and processing, capable of being stored for long periods of time and being processed far from their production site. Some studies have shown attempts to characterize the flour of grains for gluten-free bread. Perdon & Juliano [94], in comparing the effect of the amylose content and the storage of rice in the manufacturing of a fermented cake-like bread, verified that in flour where the concentration of amylose was 20-25%, the product presented higher volume and excellent softness when compared to flour with high or low amylose content, which resulted in the bread's collapse during baking. Nishita & Bean [83] confirmed these results.

Nishita & Bean [83], when analyzing the physico-chemical properties of various kinds of rice flour, verified that rice flour obtained from medium-sized rice grains, with amylose content close to 20% and low gelitanization temperature (60%), viscosity at 95°C at 750 BU (Brabender Units) and viscosity at the end of the cooling cycle at 50°C at 715 BU achieved better results in the manufacturing of rice bread, when compared to other kinds of flour that had characteristics different than these.

However, tubers and roots have a low yield and are very perishable due to their high water content. They must be processed close to their production site [31] and it is recommended that they be used after flour or starch processing.

Kim & De Ruiter [69, 70] considered that tuber and root flours are difficult to characterize, to the point where they studied various parameters during the production of cassava-soy bread and concluded that:

- Cassava-soy dough is a semi-liquid batter, which lacks cohesive and elastic properties. This bread had to be produced with cake-making equipment rather than the traditional bread-making equipment
- Cassava flour showed much greater variations in quality when compared to cassava starch. Other variations such as cassava starch may be substituted for corn starch or the mixture of corn-cassava starch
- The speed of the mixture influenced the texture of the bread's crumb, which became finer and softer when the mixing was faster
- The time for fermentation was from 40 to 68 minutes and had little effect on the bread's volume. The breadcrumb had a thicker texture, but when an intermediary fermentation phase was added to the process, the crumb had a finer and softer texture
- The increase in sugar concentration from 0, 1, or 2 to 4% resulted in a breadcrumb with a finer and softer texture, and a lighter color. The color of the crust became progressively darker and the bread was sweeter in flavor

The processing of gluten-free bread is similar to that of cake manufacture, although with yeast-fermentation similar to that of wheat bread, because the dough of gluten-free bread does not have viscoelastic properties and must therefore be manufactured with cake-making equipment [21, 74, 83, 84]. Additionally, since the dough is in a pasty format, it cannot be molded and must be placed into its mold with a spoon [83, 84].

Additional studies have been made, such as Sikora [117], which measured the rheological properties of gluten-free dough (GFD) as well as conventional gluten containing dough (CGD) during processing. This was done in order to determine the gelatinization temperature of starches included in the dough during the baking process.

Leray [73] studied the effects of freezing and frozen storage conditions of four dough formulations: standard wheat dough, fiber-enriched wheat dough, standard gluten-free dough and gluten-free dough containing amaranth flour, without use of yeast in formulations. He determined that in gluten-free dough, the addition of amaranth flour to gluten-free dough also increases its resistance to freezing but decreases its resistance to storage condition. The possibility of freezing the dough can facilitate the domestic production of gluten-free bread.

The baking properties of the pseudo cereals amaranth, quinoa and buckwheat as potential healthy and high-quality ingredients in gluten-free breads [4, 5, 6, 18] were investigated by Alvarez-Jubete [4, 5, 6], in which they uncovered the following:

- Bread volumes were significantly higher for the buckwheat and quinoa breads in comparison with the control sample

- Pseudo cereal-containing breads were characterized by a significantly softer crumb texture effect that was attributed to the presence of natural emulsifiers in the pseudo cereal flours;
- No significant differences were obtained in the acceptability of the pseudo cereal-containing gluten-free breads in comparison with the control sample

Capriles et al. [18] developed gluten-free breads with amaranth, corn and fructans; they obtained bread with reduced glycemic response and high consumer sensory acceptability. Gambuś et al [50] obtained the same results for gluten-free rolls with amaranth.

Investments made in gluten-free bread production have been few but have recently been increasing, due to the extensive broadcast about celiac disease, supported by better patient diagnosis, and by associations forming all over the world that focus on providing a better quality of life for celiac patients [2, 48, 59, 104, 114].

At present, research and development of gluten-free bread is underway, attempting to increase the supply of these products in the diet of celiac and wheat-allergic patients.

In order to improve the quality of gluten-free bread (GFB), many researchers have used various additives. One of the most frequently used are gums, emulsifiers and gelatinized flour or starch [29, 47, 84, 126].

5.1. Additives and Ingredients Used in the Production of Gluten-Free Bread

The main objectives of a new ingredient or additive study for gluten-free bread are to:

- Improve the specific volume of the bread (approx. 2-2,5 mLg^{-1}), which is low in comparison to wheat bread (approx. 5-6 mLg^{-1}) [21, 21, 23, 24, 25]
- Improve the bread's texture, which is very firm after cooling [22, 23, 24, 24]
- Extend the shelf life of gluten-free bread through decreased starch retrogradation [22, 24, 25]

Initially, the search for agents that would stimulate water retention and lower starch retrogradation was the focus of most research, and therefore the use of gums and gelatinized starch was tested by many researchers [21, 22, 23, 25, 28, 47, 68, 69, 70, 74, 83, 84, 85, 86, 123, 131]. Fats and emulsifiers began to be used in attempt to improve the texture of bread that showed excessive firmness [29, 65, 88, 131]. Later endeavors presented protein supplements such as soy protein and egg albumen, in an attempt to form a three-dimensional net capable of retaining gas. Research is progressing and showing promising results regarding the use of enzymes that promote oxidation, such as glucose oxidase [56], and reactions between groups of amino acids, such as transglutaminase [79, 119].

In 1954, Rotsch [110] showed that it is possible to prepare bread from starches that do not contain gluten-forming proteins, as long as a swelling agent is added during dough preparation.

Jongh [65] determined that dough made from pure starch and water does not retain fermentation gas, hence part of the gas escapes and the bread cells become irregular. This author suggested the use of binding agents such as fat, glyceryl monoestearate, egg whites, and gliadin. All agents improved crumb texture and loaf volume. In this case, gluten free bread was developed without the intention of supplying celiac patients.

Kim & Ruiter [69] tested twenty different swelling agents (carboxymethylcellulose, alginates, gums and pregelatinized starches) and surfactants (soybean lecithin and glyceryl monoestearate). They used composite flour of 70 parts of yam flour and 30 parts of low-fat peanut flour, and they obtained low specific volumes (1,7 to 2,0 mL/g).

FAO [47] published the possibility of manufacturing bread made with sorghum grains, manioc and other edible native starches, substituting the gluten net with gelatinized starch.

- Sorghum bread made with dry xanthan was dense and compact, but the properties were better than with the use of hydrated xanthan. Both kinds of bread were ready for slicing within three hours of baking and they kept moist and fresh for at least six days
- Cassava breads also tasted delicious and had excellent texture, although the loaves were gummy and wet when first taken out of the oven. They were ready for slicing after about 20 hours and stayed fresh and moist for eight to nine days

The gluten free bread had technological characteristics unequal to wheat, and its manufacture in a commercial scale presented problems such as texture and viscoelasticity much below the expected. Cooking the starch, in large quantities, results in a significant increase in viscosity during the heating process. This not only makes it harder to handle the paste, but it also increases the amount of energy used in the conventional process of gelatinizing it and bringing it to room temperature.

Ylimaki et al. [126] made a gluten-free wheat bread from rice flour (80%) and potato starch (20%), using response surface methodology to find a proper carboxymethylcellulose (CMC) - hydropropylmethylcellulose (HPMC) - water combination and showed that CMC and water had the greatest effect on the responses measured while HPMC had the least.

Silva et al. [118] used corn flour, cornstarch, and manioc flour to obtain hybrid "gluten-free" bread. Its was observed that pre-gelatinization, pre-fermentation and the addition of CMC improved the rheology of the corn flour dough, although the breads did not show adequate characteristics and the inclusion of other additives (Sodium di-2-stearoyl lactate and egg albumin) produced satisfactory results for hybrid breads made with native cornstarch and manioc flour.

By substituting wheat flour in a white bread formulation and using only rice flour and extruded rice flour as gluten substitutes at the level of 10% (rice flour bases), Clerici [21] and Clerici & El-Dash [22, 24] determined that rice flour extruded at high temperatures (180°C) and low moisture (20%) was found best suitable for the manufacture of gluten free bread. Under these conditions, the bread had a similar color to that of conventional wheat bread, but not the volume and texture.

Borges et al. [15] produced bread using manioc and soy with additives. Mixing must be done vigorously in order to incorporate air into the dough and retain it there; otherwise, gases easily escape the mix. The mixing should be sufficient to produce a high and constant number of air bubbles, obtaining texture similar to that of wheat bread, where few and large bubbles are signature of undesirable texture. The specific volume for this hybrid bread was 2.48 mL/g.

Defloor [28] and Defloor et al. [29] tested the effect of an extruded starch and an emulsifier (Glyceryl monostearate) on the manufacture of manioc bread made with defatted soy flour. It was verified that both additives improve gas retention during fermentation, influence the starch paste's properties, positively affect the volume and structure of the bread's centre, and are responsible for the bread's flavor. The emulsifier increased the incorporation of air during mixing; resulting in the formation of few cells of gas in the dough, while the extruded starch mainly influenced the dough's viscosity. Using the emulsifier in 4% resulted in a 30% loaf volume increase, and using the extruded starch in 9-20% increased loaf volume by 12%.

Ranhotra et al. [103] produced bread made with wheat starch fortified with soy protein isolate and verified that using the soy at 20% improved volume and internal characteristics of the bread. At 40% use results remained satisfactory but the bread's quality became worse, especially regarding the colors of the crust and crumb, which became darker. The use of 14% of sacarose produced bread with a pleasant flavor, although lacking in sweetness. The use of other ingredients in quantities above the indicated amount did not improve the bread's quality.

While developing a formulation for rice-based fermented bread, Nishita et al. [84] tested the effect of various ingredients on the quality of the final product. It was verified that hydroxipropil cellulose was the only ingredient capable of improving the dough providing enough viscosity to retain the gases produced during fermentation and through the release of enough water to gelatinize the starch during baking. Fats and emulsifiers had a negative effect, whereas refined vegetable oil produced bread loaves with satisfactory volume and texture. Through sensorial analysis of the rice bread compared with wheat bread, they observed that less than half of the panel attributed a score higher than five, on a scale of one to nine.

Casier et al. cited by Dendy [31] described a process for making wheat-less bread with sorghum and millet and pentosans insoluble in water.

Satin [113] developed a method for GFB manufacture without the use of surfactants or emulsifiers based on the fact that the gelatinization of a fraction of the flour (not wheat) increases viscosity, benefiting the structure of the dough.

In terms of volume, Akobundu et al. [3] obtained a specific volume of 2.35 mLg^{-1} for GFB's made with a mix of corn flour, cornstarch, soybean and water. DeFloor et al. [29] obtained a specific volume of 2.6 mLg^{-1} for bread made with manioc flour, soy flour and extruded cornstarch. When this last ingredient was substituted for glyceryl monostearate, the volume was 2.9 mLg^{-1}.

At present, Brazilian endeavors to develop gluten-free products has been based on the theory published by El-Dash [41], where the gelatinized starch is capable of producing a three-dimensional net formed by hydrogen bonds and has enough viscoelastic properties to retain gases produced during the bread's fermentation process.

Based on this theory and using the formula published by FAO [47], Machado [74] obtained a specific volume of 0.94 to 1.34 mLg^{-1} for breads that had their gluten substituted for gelatinized rice flour (produced through the method of cooking with water and subsequently lowering its temperature).

Clerici [21], Clerici & El-Dash [22, 23, 24] and Clerici et al. [25] tested the chemical and physical modifications of starch found in rice flour by simultaneously processing the starch through extrusion and chemical modifications (extruded phosphated rice flour, extruded cross-linked rice flour and extruded acid-modified rice flour, that were used at the level of 10% [rice flour base]). The objective was to strengthen the net of starch with ionic and covalent bonds. When producing gluten free bread, they substituted the gelatinized rice flour for the chemically modified flours, obtaining breads that had average specific volumes of 2.10 to 2.35 mLg^{-1}, as well as crust and crumb colors and textures similar to wheat bread. Results were promising and new research could be driven toward strengthening the net of starch.

5.2. Main Studies Focused on the Development of Gluten-Free Bread

5.2.1. Genetic

There is research that seeks a bread wheat line, but studies in search of a line that is exempt of a percentage of gliadin toxic to celiac patients are already underway. This is such that Carroccio et al. [19] evaluated the bread wheat line C173 for its effects on the in-vitro-grown duodenal mucosa of CD (celiac disease) patients. Tests were done on in-vitro-grown duodenal mucosa biopsies of 19 CD patients on a gluten-free diet that were exposed to peptic/tryptic-digested prolamins. Tests showed that gliadin-deficient wheat has a lower direct toxicity but activates an immunologic reaction of the duodenal mucosa like that of the common wheat species.

Genetically modifying wheat so that it does not have a fraction of gliadin (which is toxic for celiac patients), but in a way that it is also capable of forming gluten with good technological qualities for bread manufacture, could represent a mark in the area of Food Science and Technology.

5.2.2. Enzyme

The use of enzymes that trigger reactions capable of creating a three-dimensional net similar to that of gluten is currently under investigation.

The use of microbial transglutaminase (MTGase) in bakery products has been studied as an efficient strengthener of the gluten network, allowing not only the usage of weaker flours in products that require strong flours, but also the incorporation of new protein sources in those foods [119].

Moore et al. [79] studied the addition of enzymes at protein levels of 0.0, 0.1, 1.0 and 10.0 Ug^{-1}, in formulas including rice, potato and corn flours, xanthan gum and several protein sources (soy and egg). Results show that it is possible to create a protein net to be used in the manufacture of gluten-free bread using enzymes, considering that its efficiency is dependent on the quantity of added enzymes and which protein source is used.

Onyango [86] investigated the effect of different concentrations (0, 0.5, 1 and 1.5 Ug^{-1}) of microbial transglutaminase (MTG) on the creep-recovery properties of gluten-free batter prepared from pre-gelatinized cassava starch, sorghum and egg albumen. Crumb properties of gluten-free bread baked from the batter revealed that increasing MTG concentration increased crumb firmness and chewiness, whereas increasing incubation time decreased crumb cohesiveness, chewiness and resilience. There were no significant interaction effects between enzyme concentration and incubation time.

Another enzyme that is used to strengthen wheat flour is glucose oxides (GO), which has been tested in the manufacture of gluten-free bread.

GO has been used to promote the oxidation of the protein matrix, drawing considerable interest as an "improver" of wheat flour dough [34]. It catalyzes the oxidation reaction of glucose to gluconolactone in the presence of oxygen, which is converted into hydrogen peroxide. The hydrogen peroxide in turn promotes the oxidation of the SH group of two cystein residues to crosslinked S-S bonds in the gluten net (100, 101).

Gujral et al. [57] produced gluten-free bread using GO incorporated into the rice bread formula to improve bread quality, and discovered that GO brings about the cross linking of rice protein, and in consequence, the modification of the elastic and viscous behavior of the rice dough. The greatest benefit was that in the presence of GO the levels of HPMC required to produce acceptable rice bread could be lowered, economizing the process.

Renzetti [107] used commercial preparations of lactase (LAC) and glucose oxidase (GO) (0.01% and 0.1% addition levels), as well as protease (PR) (0.001% and 0.01% addition levels) when testing for their impact on the bread-making performance of gluten-free oat flour. LAC 0.1%, PR 0.001% and PR 0.01% additions significantly improved oat bread quality as they increased specific volume and decreased crumb hardness and chewiness. In contrast, GO 0.1% addition revealed detrimental effects, as it resulted in the hardest breadcrumb.

Gujral [55] showed the potential use of cyclodextrin glycosyl transferase (CGTase) as a rice bread improver and verified that the addition of CGTase produced a reduction in the dough consistency and the rice bread quality was better in terms of specific volume, shape index, and crumb texture.

5.2.3. Protein

Sciarini [115] used the inactive soy flour incorporation to improve bread quality parameters in both corn- and rice-based breads: volume, crumb appearance, texture, and staling and the principle results were:

- Specific volume (rice breads: 1.98 cm^3/g; rice/soy 90:10 2.51 cm^3/g, corn/soy 90:10: 2.05 cm^3/g, whereas corn/soy 80:20: 2.12 cm^3/g)
- The staling rate was decreased by soy flour incorporation on rice (staling rate of rice breads with 10% soy diminished 52%, and with 20% of soy addition, 77%, both regarding to 100% rice breads) and corn formulation (the staling rate of corn/soy 80:20 breads was 5.9% lower than corn/soy 90:10)

Miñarro [78] used unicellular protein to make starch-based and starch-vegetable-based formulations for bread. Sensorial analyses showed that starch-vegetable-based formulations without unicellular protein were the most preferred by consumers, followed by starch-vegetable-based formulations with added protein, and main differences detected by consumers were related to texture attributes (high firmness).

Milde [77] obtained optimum tapioca starch-based breads with spongy crumb, high volume and a good sensory acceptance by using the highest levels of fat and soybean flour and one egg. The bread presented low values of firmness (≤ 100 N) and elasticity (>65%) and the lowest variation of these parameters with storage, and acceptability of this bread was 84% for habitual consumers of wheat bread and 100% by celiac patients.

Pagliarini [89] determined that in the case of celiac patients, sensorial evaluation of bakery products must be based on the following sensory descriptors: porosity, crust and crumb color, softness by hand and mouth, cheese odor, corn odor and fermented odor, sweet, salty, adhesive and rubbery.

Pasqualone [91] evaluated the effectiveness of the use of cassava flour in bread making, and the sensory acceptability of the final product. Cassava breads containing egg albumen and extra-virgin olive oil, in consideration of their high nutritional value with respect to other food additives (i.e. hydrocolloids), presented significant ($p < 0.05$) improvements of loaf specific volume (from 2.24 to 3.93 mLg^{-1}) and crumb firmness (from 9.14 to 4.67 N), obtained the best scores from panelists for all the test breads examined, and resulted in a final product as attractive as the wheat bread prepared as reference.

Krupa [72] used bean starch, either native or hydrothermally modified, as an additive to a gluten-free formulation and to control a rapid staling during storage. It was determined that the addition of native starch increased the tendency of amylopectine to retrograde during storage, whereas the presence of modified starch decreased the retro gradation enthalpy by 16%, which also reduced the hardness and diminished the differences between the upper and the bottom parts of a bread slice.

Witczak [125] studied the effect of maltodextrins of various dextrose equivalents (DE) in the reduction of starch retro gradation in gluten-free bread. He concluded that the addition of maltodextrins with low DE (3.6) diminishes loaf volume and causes deterioration of bread quality. Maltodextrins with higher DE, especially 18.0 and 21.8, positively influence bread volume and have a beneficial influence on crumb hardening during storage, and is also an effective factor in reducing recrystallisation enthalpy of amylopectin.

Demirkesen [30] studied the quality of rice breads (volume, firmness and sensory analysis) obtained with different gums (xanthan gum, guar gum, locust bean gum (LBG), hydroxyl propyl methyl cellulose (HPMC), pectin, xanthan-guar, and xanthan-LBG blend) and emulsifiers (Purawave™ and DATEM) and verified that the addition of DATEM improved bread quality in terms of specific volume and sensory values.

Onyango[85] studied the effect of egg albumen, skim milk powder, soy protein isolate and soy protein concentrate on creep-recovery parameters of gluten-free batter made from sorghum and pre-gelatinized cassava starch and verified that egg albumen did not decrease bread volume and exhibited the lowest crumb firmness and staling rate. The optimization of the amount of egg albumen with diacetyl tartaric acid esters of mono and diglycerides (DATEM) showed that the use of 6% and 0.1% w/w fwb egg albumen and DATEM, respectively, resulted in gluten-free batter with the least elastic portion of maximum creep compliance ($Je/Jmax = 11.65\%$) which corresponded to the lowest crumb firmness (790.8 g).

Onyango [87] studied the effect of cellulose-derivatives and emulsifiers on the creep-recovery behavior of gluten-free dough prepared from gelatinized cassava starch and sorghum. When egg albumen powder was added (6.7% w/w fwb), several textural defects associated with gluten-free bread were eliminated. Cellulose-derivatives did not decrease crumb firmness or staling rate when compared to the control. Though increasing emulsifier concentration (from 0.4% to 2.4% w/w fwb) decreased crumb firmness, crumbs treated with 2.4% w/w fwb emulsifiers, except diacetyl tartaric acid esters of mono- and diglycerides, were weak and difficult to handle after slicing. Nevertheless, all gluten-free breads treated with 2.4% w/w fwb emulsifiers staled at a slower rate than the control.

5.2.4. Other Research

Research on this subject is diverse, varying from the addition of fiber [111], to antioxidant components and to sourdough [49, 99], *vinal* seed [10] and iron compounds [71], in the formulation of gluten-free breads. Innovative proposals have been made that can contribute to the improvement of the technological and nutritional quality of bread [129]. The main studies are as follows:

Mariotti [75] determined that *Psyllium* fiber could be used as a thickening agent and fiber source, since it forms a film-like structure that is promising for the production of gluten free bread. Zandonadi [130] confirmed that products with Psyllium fiber were well accepted by both individuals with celiac disease and individuals without celiac disease.

Moroni [80] determined that the micro biota of gluten-free dough indicates similarities with the micro biota of wheat/rye fermentation. He also suggested that the positive metabolic activities of the sourdough micro biota are still retained during fermentation of GF crops and pointed out that the use of sourdough in GF baking may be the new frontier for improving the quality, safety and acceptability of GF bread.

Torbica [122] studied the use of husked and unhusked buckwheat flour as a substitute in 10, 20 and 30% of rice flour in gluten free bread, and concluded that all six combinations of tested gluten-free breads were sensory acceptable.

According to Alvarez-Jubete [6], quinoa and buckwheat seeds and sprouts represent potential rich sources of polyphenol compounds for enhancing the nutritive properties of foods such as gluten-free breads.

Sun-Waterhouse [120] determined that the aqueous extract of kiwifruit puree containing health-beneficial constituents could be considered a functional ingredient for gluten-free bread formulation, since the aqueous extract showed good stability. High phenolic compounds and vitamin C in kiwifruit extract-enhanced bread was acceptable to a taste panel, possessing a softer and smoother texture than plain gluten-free bread.

Diowksz [33] determined that the use of 20% of inulin into lupine fiber in the basic mixture consisting of wheat, maize, and potato starches and of maize flour significantly improved the volume of bread.

Sabanis [112], studied different cereal fibers (wheat, maize, oat and barley) which were added at 3, 6 and 9 g/100 g levels into a gluten-free bread formulation based on corn starch, rice flour and hydroxypropyl methyl cellulose (HPMC). The formulation containing barley fiber produced loaves that had more intense color and volume comparable to the control, but during the storage of the breads a reduction in crumb moisture content and an increase in firmness were observed. The micrographs of the crumb showed the continuous matrix between starch and maize and/or oat fiber obtaining a more aerated structure.

Zhang & Lugar [131] patented a gluten-free dough comprising 35 to 65 wt % gluten free flour; leavening agents selected from the group consisting of 1 to 6 wt % yeast, 0.1 to 1.5 wt % chemical leavener and combinations thereof, 1 to 5 wt % modified starch; 0.05 to 1.5 wt % guar gum; 0.05 to 1.5% of xanthan gum; 0.05 to 1.2% of gum Arabic or semi-synthetic hydrophilic colloids; 30 to 55 wt % water; 2 to 20 wt % proteins; 1 to 8% sweeteners, 0.05 to 1.2 wt % emulsifiers and 1 to 15 wt % vegetable oil. According to these authors, the formulations and methods described in the patent can be used for the production of pizza crust, tortillas, bread, flat bread, biscuits, rolls and the like.

6. FUTURE PROSPECTS

The major technological problem in gluten-free bread production is to obtain a gluten substitute that is low cost and is also capable of retaining gas produced during fermentation and baking.

The studies presented in this chapter brought forth many proposals that may help in the development of gluten-free bread with good technological, sensorial, and nutritional properties, especially since:

- A gluten substitute can be obtained through starches, chemically modified starches, proteins, gums and fibers
- The use of enzymes that promote the formation of a strong three-dimensional protein net has already proven itself to be efficient in the case of weak wheat flours. The use of these enzymes has already been tested for other proteins that are candidates for gluten substitutes
- The use of emulsifiers, gums, and modified starches has appeared promising in terms of halting starch retro gradation during storage and can be use in products' shelf-life study
- Texture, color, and flavor characteristics of gluten-free breads are already being well accepted by panelists, indicating that new formulations are coming closer to gluten-free bread with good technological qualities

In addition to future prospects for gluten-free products, the advances in other areas are already very promising, such as:

- Genetics is already showing us that it will soon be possible to obtain wheat without a fraction of the protein that triggers celiac disease, but that is also capable of producing bread with good volume and texture
- In medicine, the study of enzymes destined for the development of enzymatic digestion of gluten in the intestine has shown itself to be promising in the case of Prolyl endopeptidase from *Myxococcus xanthus* (MX PEP) (51) and Sunn pests (*Eurygaster integriceps*) [27]

CONCLUSION

The quality of gluten-free bread (GFB) is not yet similar to that of wheat bread, but the results obtained at present are promising and make way for new investigations that may contribute to increasing the supply of bakery products for celiac patients. As the quality of GFB improves, other products (such as cakes, pizza, and cookies) will reap the benefits of new discoveries.

REFERENCES

[1] AACC - American Association of Cereal Chemists. (2000). Approved methods. 10 ed. American Association of Cereal Chemists, Saint Paul, Minnesota.

[2] ACELBRA-Associação dos Celíacos do Brasil. (2008, may, 12). Available in: <http://www.acelbra.org.br/2004/estatisticas.php>.

[3] Akobundu, C. N., Ubbaonu, C. N., & Ndupuh, C. E. (1988). Studies on the baking potencial of non-wheat composite flours. *Journal of Food Science and Technology, 25, 4,* 211-214.

[4] Alvarez-Jubete, L., Arendt, E. K., & Gallagher, E. (2009). Nutritive value and chemical composition of pseudocereals as gluten-free ingredients. *International Journal of Food Sciences and Nutrition, 60, 4,* 240-257.

[5] Alvarez-Jubete, L., Auty, M., Arendt, E.K., & Gallagher, E. (2009). Baking properties and microstructure of pseudocereal flours in gluten-free bread formulations. *European Food Research and Technology, 230, 3,* 437-445.

[6] Alvarez-Jubete, L., Wijngaard, H., Arendt, E. K., & Gallagher, E. (2010). Polyphenol composition and in vitro antioxidant activity of amaranth, quinoa buckwheat and wheat as affected by sprouting and baking. *Food Chemistry, 119, 2,* 770-778.

[7] Bauer, N., Koehler, P., Wieser, H., & Schieberle, P. (2003). Studies on Effects of Microbial Transglutaminase on Gluten Proteins of Wheat. II. Rheological Properties. *Cereal Chemistry, 80, 6,* 787-790.

[8] Beckwith, A.C., & Wall, J.S. (1966). Reduction and reoxidation of wheat glutenin. *Biochimica et Biophysica Acta, 13,* 155-162.

[9] Beckwith, A.C., Wall, J.S., & Dimler, R.J. (1965). Amide groups and interaction sites in wheat gluten proteins, part I: effects of amide-ester conversion. *Archives of Biochemistry and Biophysics, 103,* 319-330.

[10] Bernardi,C., Sánchez,H., Freyre, M., & Osella, C.(2010). Gluten-free bread formulated with *Prosopis ruscifolia* (vinal) seed and corn flours. *International Journal of Food Sciences and Nutrition, 61, 3,* 245-255.

[11] Bernardin, J.E., & Kasarda, D.D. (1973). Hydrated protein fibrils from wheat endosperm. *Cereal Chemistry, 50,* 529-536.

[12] Bernardin, J. E., & Kasarda, D. D. (1973). The microstructure of wheat protein fibrils. *Cereal Chemistry, 50,* 735-745.

[13] Biagi, F., & Corazza, G.R. (2010). Mortality in celiac disease. *Nature Reviews Gastroenterology and Hepatology, 7, 3,* 158-162.

[14] Bonet, A., Rosell, C. M., Caballero, P. A., Gómez, M., Pérez-Munuera, I., & Lluch, M. A. (2006). Glucose oxidase effect on dough rheology and bread quality: A study from macroscopic to molecular level. *Food Chemistry, 99, 2,* 408-415.

[15] Borges, J. M., Ferreira, E., & Alvim, C. M. F. (1984). *Pão de mandioca e soja.* Viçosa, Universidade Federal de Viçosa.

[16] BRASIL. (1992). ANVISA, Lei 8543, December 23, 1992. Available from URL: www.anvisa.gov.br.

[17] Camargo, C. (1995). Development program for new variety of wheat in Brazil. In: *International Symposium on Food extrusion: pasta and extruded products, I*, Campinas, abstracts, 24-25.

[18] Capriles, V.D., Coelho, K.D., Matias, A.C.G., & Arêas, J.A.G. (2006). Efeito da adição de amaranto na composição e na aceitabilidade do biscoito tipo cookie e do pão de forma. *Alimentos & Nutrição, 17, 3*, 269-274.

[19] Carroccio, A., Di Prima, L., Noto, D., Fayer, F., Ambrosiano, G., Villanacci, V., Lammers, K., Lafiandra, D., De Ambrogio, E., Di Fede, G., Iacono, G., & Pogna, N. Searching for wheat plants with low toxicity in celiac disease: Between direct toxicity and immunologic activation. *Digestive and Liver Disease*. Article in Press. Doi:10.1016/j.dld.2010.05.005.

[20] Cawthorn, D. M., Steinman, H. A., & Witthuhn, R. C. (2010). Wheat and gluten in South African food products. *Food and Agricultural Immunology, 21, 2*, 91-102.

[21] Clerici, M.T.P.S. (1997). *Efeito de modificações fosfatada, intercruzada e ácida durante a gelatinização por extrusão da farinha de arroz e sua influência na produção de pão sem glúten*. Tese de Doutorado. Universidade Estadual de Campinas, UNICAMP, Campinas-SP, Brasil.

[22] Clerici, M.T.P.S., & El-Dash, A.A. (1994). Extruded rice starch as substitute for gluten in wheatless bread production. *Institute of Food Technologists Annual Meeting*, 25-29.

[23] Clerici, M.T.P.S., & El-Dash, A.A. (1995). Extruded acid-modified rice flour as substitute for gluten in wheatless bread production. *International Symposium on Food Extrusion: Pasta and Extruded Products*, Campinas, Brazil, March 27-29.

[24] Clerici, M.T.P.S., & El-Dash, A.A. (2006). Farinha extrusada de arroz como substituto de glúten na produção de pão de arroz. *Archivos Latinoamericanos de Nutrición, 56, 3*, 288-294.

[25] Clerici, M. T. P. S., Airoldi, C., & El-Dash, A. A. (2009). Production of acidic extruded rice flour and its influence on the qualities of gluten-free bread. *LWT - Food Science and Technology, 42, 2*, 618 - 623.

[26] Cornell, H. J., & Hoveling, A. W. (1998). *Wheat: Chemistry and Utilization*. Lancaster: Technomic Publishing Co., Inc.

[27] Darkoh, C. (2008). *Isolation, purification, and characterization of gluten-specific enzyme from Sunn pest, Eurygaster integriceps* [M.S. dissertation]. United States - Texas: Stephen F. Austin State University; In: Dissertations & Theses: A&I [database on the Internet] [cited 2010 Jun 24]. Available from: http://www.proquest.com/; Publication Number: AAT 1459724.

[28] Defloor, I. (1995). *Factors governing the breadmaking potential of cassava (Manihot esculenta Crantz) flour* [Ph.D. dissertation]. Belgium: Katholieke Universiteit Leuven (Belgium); In: Dissertations & Theses: A&I [database on the Internet] [cited 2010 Jun 24]. Available from: http://www.proquest.com/; Publication Number: AAT C446516.

[29] Defloor, I., De Geest, C., Schllekens, M., Martens, A., & Delcour, J.A. (1991). Emulsifiers and/or extruded starch in the production of breads from cassava. *Cereal Chemistry, 68, 4*, 323-7.

[30] Demirkesen, I., Mert, B., Sumnu, G., & Sahin, S. (2010). Rheological properties of gluten-free bread formulations. *Journal of Food Engineering, 96, 2*, 295-303.

[31] Dendy, D. (1992). Perspectives in composite and alternative flour products. In. Cereal Chemistry and Technology: a long past and bright future. *9th International Cereal and Bread Congress*, Paris, 1-5, June.

[32] DeStefanis, V. (1994). *Benzoyl peroxide to improve the performance of oxidants in breadmaking*. Patent 5318785, Issued on June 7, 1994.

[33] Diowksz, A., Sucharzewska, D., & Ambroziak, W. (2009). Function of dietary fibre in forming functional properties of gluten free dough and bread. *Zywnosc. Nauka. Technologia. Jakosc/Food Science Technology Quality, 16, 2,* 83-93.

[34] Dunnewind, B., Van Vliet, T., & Orsel, R. (2002).Effect of oxidative enzymes on bulk rheological properties of wheat flour doughs. *Journal of Cereal Science, 36,* 357-366.

[35] El-Dash, A.A. (1978). Standardized mixing and fermentation procedure for experimental baking test. *Cereal Chemistry, 55, 4,* 436-446.

[36] El-Dash, A. A. (1991). Molecular structure of gluten and viscoelastic properties of dough: a new concept. *Proceedings of the First Brazilian Congress Proteins.* Ed. UNICAMP, 513-530.

[37] El-Dash, A. A., Camargo, C. R. O., & Diaz, N. (1982). *Fundamentos de Tecnologia de Panificação.* São Paulo, Série Agro-Industrial.

[38] El-Dash, A., & Germani, R. (1994). *Tecnologia de farinhas mistas: uso de farinha mista de trigo e milho na produção de pães.* Brasília, EMBRAPA - SPI.

[39] El-Dash, A., Cabral, L.C., & Germani, R. (1994). *Uso de farinha mista de trigo e soja na produção de pães.* Brasília, EMBRAPA - SPIPP.

[40] El-Dash, A., Campos, J. E., & Germani, R. (1994). *Tecnologia de farinhas mistas: uso de farinha mista de trigo e sorgo na produção de pães.* Brasília, EMBRAPA - SPIPP.

[41] El-Dash, A., Mazzari, M. R., & Germani, R. (1994). *Tecnologia de farinhas mistas: uso de farinha mista de trigo e mandioca na produção de pães.* Brasília, EMBRAPA - SPIPP.

[42] Elson, C.O. (2005). National institutes of health consensus development conference statement on celiac disease. *Gastroenterology, 128, 4,* S1-S9.

[43] Ewart, J. A. D. (1968). A hypothesis for the structure and rheology of glutenin. *Journal of the Science of Food and Agriculture, 19,* 617-623.

[44] Ewart, J. A. D. (1972). A modified Hypothesis for the structure and Rheology of Glutelins. *Journal of the Science of Food and Agriculture, 23,* 687-699.

[45] Ewart, J. A. D. (1979). Glutenin structure. *Journal of the Science of Food and Agriculture, 30,* 482-492.

[46] FAO-Food Outlook-November 2008. Wheat. Available from: http://www.fao.org/docrep/011/ai474e/ai474e03.htm.

[47] FAO. (1989). *Wheatless bread.* Rome. January.

[48] FENACELBRA (2007). Available from: http://www.doencaceliaca.com.br/index.htm.

[49] Galle, S., Schwab, C., Arendt, E., & Gänzle, M. (2010). Exopolysaccharide-forming weissella strains as starter cultures for sorghum and wheat sourdoughs. *Journal of Agricultural and Food Chemistry, 58, 9,* 5834-5841.

[50] Gambuś, H., Gambuś, F., Wrona, P., Pastuszka, D., Ziobro, R., Nowotna, A., Kopeć, A., & Sikora, M. (2009). Enrichment of gluten-free rolls with amaranth and flaxseed increases the concentration of calcium and phosphorus in the bones of rats. *Polish Journal of Food and Nutrition Sciences, 59, 4,* 349-355.

[51] Gass, J. D. (2007). *Oral enzyme therapies for Celiac Sprue* [Ph.D. dissertation]. United States - California: Stanford University; In: Dissertations & Theses: A&I [database on the Internet] [cited 2010 Jun 24]. Available from: http://www.proquest.com/; Publication Number: AAT 3253483.

[52] Goldstein, S. (1957). Sulfhydryl und Disulfidgruppen der Klebereiweisse and ihre Beziehung zur Backfahigkeit der Brotmehle. *Mitt. Gebiete Lebensmittelunters. Hyg., 48*, 87-93.

[53] Goñi, I., Garcia-Alonso, A., & Saura-Calixto, F. (1997). A starch hydrolysis procedure to estimate glycemic index. *Nutrition Research, 17, 3*, 427-437.

[54] Greenwood, G.T., & Ewart, J.A.D. 1975. Hypothesis for the structure of glutenin in relation to rheological properties of gluten and dough. *Cereal Chemistry, 52, 2*, 146-153.

[55] Gujral, H. S. (2003). Effect of Cyclodextrinase on Dough Rheology and Bread Quality from Rice Flour. *Journal of Agricultural and Food Chemistry, 51, 13*, 3814-3818.

[56] Gujral, H. S., & Rosell, C. M. (2004). Improvement of the breadmaking quality of rice flour by glucose oxidase. *Food Research International, 37, 1*, 75–81.

[57] Gujral, H.S., Gaur, S., & Rosell, C. M. (2003). Note: Effect of Barley Flour, Wet Gluten and Ascorbic Acid on Bread Crumb Texture. *Food Science and Technology International, 9*, 17-21.

[58] Hadjivassiliou,M., Sanders, D. S., Grünewald, R. A., Woodroofe, N., Boscolo, S., & Aeschlimann, D. (2010). Gluten sensitivity: from gut to brain. *The Lancet Neurology, 9, 3*, 318 – 330.

[59] Hoggan, R., & Adams, S. Cereal Killers: Celiac Disease and Gluten-Free A to Z. http://www.celiac.com.

[60] Holm,K., Maki, M., Vuolteenaho,N., Mustalahti,K., Ashorn,M., Ruusca, T., & Kaukinens,K. (2006). Oats in the treatment of childhood celiac desease: a 2-year controlled trial and a long-term clinical follow up study. *Alimentary Pharmacology & Therapeutics, 23, 10*, 1463-1472.

[61] Houben, A., Götz, H., Mitzscherling, M., & Becker, T. (2010). Modification of the rheological behavior of amaranth (*Amaranthus hypochondriacus*) dough. *Journal of Cereal Science, 51, 3*, 350-356,

[62] Howdle,P. (2002). Coeliac disease: a clinical update. In: Emerton, V. *Food Allergy and intolerance-current issues and concerns*. Cambridge, UK: Food Ra leatherhead, 114-126.

[63] Hüttner, E. K., & Arendt, E. K. (2010). Recent advances in gluten-free baking and the current status of oats. *Trends in Food Science and Technology, 21,6* ,303-312.

[64] Jarmo, K., Visakorpi, J., & Mäki, M. (1995). Variaciones en las características clínicas de la enfermidad celíaca. *Alimentaria, 33, 264*, 93-96.

[65] Jongh, G. (1961) .The formation of dough and bread structures. I. The ability of starch to form structures, and the improving effect of glyceryl monostearate. *Cereal Chemistry, 38*, 140-152.

[66] Kagnoff, M.F. (1985). Coeliac disease: genetic, immunological and environmental factors in disease pathogenesis. *Scandinavian Journal of Gastroenterology - Supplement, 20*, S114, 45-54.

[67] Kagnoff, M.F. (2005). Overview and pathogenesis of celiac disease. *Gastroenterology, 128, 4*, S10-S18.

[68] Kasarda, D.D., Bernardin, J.E., & Nimmo, C.C. (1976). Wheat proteins. In: Pomeranz, Y. *Advanced Cereal Science and Technology*, American Association of Cereal Chemistry.

[69] Kim, J. C., & De Ruiter, D. (1968). Bread from non-wheat flours. *Food Technology, 22,* 867-878.

[70] Kim, J. C., & De Ruiter, D. (1969). Bakery products with non-wheat flours- a review. *Bakers Digest, 43, 3,* 58-63.

[71] Kiskini, A., Kapsokefalou, M., Yanniotis, S., & Mandala, I. (2010). Effect of different iron compounds on wheat and gluten-free breads. *Journal of the Science of Food and Agriculture, 90, 7,* 1136-1145.

[72] Krupa, U., Rosell, C. M., Sadowska, J., Soral-ŚMietana, M. (2010). Bean starch as ingredient for gluten-free bread. *Journal of Food Processing and Preservation, 34, 2,* 501-518.

[73] Leray, G., Oliete, B., Mezaize, S., Chevallier, S., & Lamballerie, M. (2010). Effects of freezing and frozen storage conditions on the rheological properties of different formulations of non-yeasted wheat and gluten-free bread dough, *Journal of Food Engineering, 100, 1,* 70-76.

[74] Machado, L. M. P. (1996). *Pão sem glúten: otimização de algumas variáveis de processamento.* Campinas. Dissertations of Master Science. Faculdade de Engenharia de AlimentosUniversidade Estadual de Campinas.

[75] Mariotti, M., Lucisano, M., Ambrogina Pagani, M., & Ng, P.K.W. (2009).The role of corn starch, amaranth flour, pea isolate, and Psyllium flour on the rheological properties and the ultrastructure of gluten-free doughs. *Food Research International, 42, 8,* 963-975.

[76] Menezes, E.W., Giuntini, E.B., Dan, M.C.T., & Lajolo, F.M. (2009). New information on carbohydrates in the Brazilian Food Composition Database. *Journal of Food Composition and Analysis, 22, 5,* 446–452.

[77] Milde, L.B., Ramallo, L.A., & Puppo, M.C. (2010). Gluten-free Bread Based on Tapioca Starch: Texture and Sensory Studies. *Food and Bioprocess Technology,* 1-9. DOI 10.1007/s11947-010-0381-x.

[78] Miñarro, B., Normahomed, I., Guamis, B., & Capellas, M. (2010). Influence of unicellular protein on gluten-free bread characteristics. *European Food Research and Technology, 231, 2,* 171-179.

[79] Moore, M.M., Heinbockel, M., Dockery, P., Ulmer, H.M., Arendt, E.K. (2006). Network Formation in Gluten-Free Bread with Application of Transglutaminase. *Cereal Chemistry, 83, 1,* 28-36.

[80] Moroni, A.V., Dal Bello, F., & Arendt, E. K. (2009). Sourdough in gluten-free bread-making: An ancient technology to solve a novel issue? *Food Microbiology, 26, 7,* 676-684.

[81] Nasmith, G. G. (1903). The Chemistry of Wheat Gluten. University of Toronto, June ist,. The Transactions of the Canadian Institute. Vol. VII. Available from: http://www.archive.org/stream/chemistryofwheat00nasmuoft/chemistryofwheat00nasmuoft_djvu.txt

[82] Niewinski, M.M. (2008). Advances in celiac disease and gluten-free diet. ***Journal of the American Dietetic Association**, 108, 4,* 661-72.

[83] Nishita, K. D., & Bean, M. M. (1979). Physicochemical properties of rice in relation to rice bread. *Cereal Chemistry, 56, 3,* 185.

[84] Nishita, K. D., Roberts, R. L., & Bean, M. M. (1976). Development of a yeast-leavened rice-bread formula. *Cereal Chemistry, 53, 5,* 626-635.

[85] Onyango, C., Mutungi, C., Unbehend, G., & Lindhauer, M. G. (2009). Creep-recovery parameters of gluten-free batter and crumb properties of bread prepared from pregelatinised cassava starch, sorghum and selected proteins. *International Journal of Food Science and Technology, 44, 12,* 2493-2499.

[86] Onyango, C., Mutungi, C., Unbehend, G., & Lindhauer, M. G. (2010). Rheological and baking characteristics of batter and bread prepared from pregelatinised cassava starch and sorghum and modified using microbial transglutaminase, *Journal of Food Engineering, 97, 4,* 465-470.

[87] Onyango, C., Unbehend, G., & Lindhauer, M. G. (2009). Effect of cellulose-derivatives and emulsifiers on creep-recovery and crumb properties of gluten-free bread prepared from sorghum and gelatinised cassava starch. *Food Research International, 42, 8,* 949-955.

[88] Özer, M. S., Kola, O., & Duran, H. (2010). Effects of buckwheat flour combining phospholipase or DATEM on dough properties *Journal of Food, Agriculture and Environment, 8, 2,* 13-16.

[89] Pagliarini, E., Laureati, M., & Lavelli, V. (2010). Sensory evaluation of gluten-free breads assessed by a trained panel of celiac assessors. *European Food Research and Technology, 231, 1,* 37-46.

[90] Palosuo, K., Varjonen, E., Nurkkala, J., Kalkkinen, N., Harvima, R., Reunala, T., & Alenius, H. transglutaminase-mediated cross-linking of a peptic fraction of ω-5 gliadin enhances IgE reactivity in wheat dependent, exercise-induced anaphylaxis. ***Journal of Allergy and Clinical Immunology, 111,6,*** 1386-1392.

[91] Pasqualone, A., Caponio, F., Summo, C., Paradiso, V. M., Bottega, G., & Pagani, M. A. (2010). Gluten-free bread making trials from cassava (Manihot Esculenta Crantz) flour and sensory evaluation of the final product. *International Journal of Food Properties, 13, 3,* 562-573.

[92] Peña, R.J., Amaya, A., Rajaram, S., & Mujeeb-Kazi, A. (1990). Variation in quality characteristics associated with some spring 1B/1R translocation wheats. *Journal of Cereal Science, 12, 2,* 105-112.

[93] Peräaho, M. P., Collin, P., Kaukinen, K., Kekkonen, L., Miettinen, S., & Mäki, M. (2004). Oats Can Diversify a Gluten-Free Diet in Celiac Disease and Dermatitis Herpetiformis. *Journal of the American Dietetic Association,104,7,* 1148-1150.

[94] Perdon, A. A., & Juliano, B. O. (1975). Amylose content of rice and quality of fermented cake. *Die Stärke, 27, 6,* 196-198.

[95] Pietzak, M. (2005). Follow-up of patients with celiac disease: Achieving compliance with treatment. *Gastroenterology, 128,* S135-S141

[96] Pietzak, M., Catassi, C., Drago, S., Fornaroli, F., & Fasano, A. (2001). Celiac disease: Going against the grains. *Nutrition in Clinical Practice, 16,* 335-344.

[97] Polanco, I., Molina, M., Pietro, G., Carraco, S., & Lama, R. (1995). Dieta y enfermidad celíaca. *Alimentaria, 33, 264,* 91-93.

[98] Pomeranz, Y. (1987). *Modern Cereal Science and Technology*, VCH Publishers, New York.

[99] Poutanen, K., Flander, L., & Katina, K. (2009). Sourdough and cereal fermentation in a nutritional perspective. *Food Microbiology, 26, 7,* 693-699.

[100] Primo-Martín, C., & Martínez-Anaya, M.A. (2003). Influence of pentosanase and oxidases on water-extractable pentosans during a straight breadmaking process. *Journal of Food Science, 68, 1*, 31-34.

[101] Primo-Martín, C., Wang, M., Lichtendonk, W.J., Plijter, J.J., Hamer, R.J. (2005). An explanation for the combined effect of xylanase - glucose oxidase in dough systems. *Journal of the Science of Food and Agriculture, 85, 7,* 1186-1196.

[102] Pyler, E. J. (1988). *Baking - Science e Tecnology.* 3.ed. vol II. Sosland Publishing Company, Kansas.

[103] Ranhotra, G. S., Loewe, R. J., & Puyat, L. V. (1975). Preparation and evaluation of soy-fortified gluten-free bread. *Journal of Food Science, 40, 1,* 62-64.

[104] Rauen, M. S., Back, J. C. V., & Moreira, E. A. M. (2005) Celiac disease's relationship with the oral health. *Revista de Nutrição, 18, 2,* 271-276.

[105] Redman, D. G., & Ewart, J. A. D. (1967). Disulphide interchange in dough proteins. *Journal of the Science of Food and Agriculture, 18, 1,* 15-18.

[106] Redman, D. G., & Ewart, J. A. D. (1967). Disulphide interchange in cereal proteins. *Journal of the Science of Food and Agriculture, 18, 11,* 520-523.

[107] Renzetti, S., Courtin, C. M., Delcour, J. A., & Arendt, E. K. (2010). Oxidative and proteolytic enzyme preparations as promising improvers for oat bread formulations: Rheological, biochemical and microstructural background. *Food Chemistry, 119, 4,* 1465-1473.

[108] Roberts, S.B. (2000). High-glycemic index foods, hunger, and obesity: is there a connection? *Nutrition Reviews, 58,* 163-169.

[109] Robertson, G. H., Gregorski, K. S., & Cao, T. K. (2006). Changes in secondary protein structures during mixing development of high absorption (90%) flour and water mixtures. *Cereal Chemistry, 83, 2,* 136–142.

[110] Rotsch, A. (1954). Chemische und backtechnische Untersuchungen an kunstlichen Teigen, *Brot u. Geback, 8,* 129-130.

[111] Sabanis, D., Lebesi, D., & Tzia, C. (2009). Development of fibre-enriched gluten-free bread: A response surface methodology study. *International Journal of Food Sciences and Nutrition, 60, 4,* 174-190.

[112] Sabanis, D., Lebesi, D., & Tzia, C. (2009). Effect of dietary fibre enrichment on selected properties of gluten-free bread. *LWT - Food Science and Technology, 42, 8,* 1380-1389.

[113] Satin, M. Bread without wheat. (1988). *New Scientist, 28, 1610,* 56-59.

[114] Schober, T. J., Scott, R. B. (2010). *Gluten-free baking: what is happening inside the bread?* United States Department of Agriculture, Agricultural Research Service, Grain. In www.csaceliacs.org/documents/Gluten-freebaking_000.pdf

[115] Sciarini, L. S., Ribotta, P. D., León, A. E., & Pérez, G. T. (2010). Influence of Gluten-free Flours and their mixtures on batter properties and bread quality. *Food and Bioprocess Technology, 3, 4,* 577-585.

[116] Setty, M., Hormaza, L., & Guandalini, S. (2008). Celiac disease: Risk assessment, diagnosis, and monitoring. *Molecular Diagnosis and Therapy, 12, 5,* 289-298.

[117] Sikora, M., Kowalski, S., Krystyjan, M., Ziobro, R., Wrona, P., Curic, D., & LeBail, A. (2010). Starch gelatinization as measured by rheological properties of the dough. *Journal of Food Engineering, 96, 4,* 505-509.

[118] Silva, A. C. Q. R., Nunes, M. L. & Silva, C. E. M. (1994). Influência de alguns parâmetros na elaboração de "pães sem glúten". In.: *Congresso Brasileiro de Ciência e Tecnologia de Alimentos, XIV,* São Paulo, Anais.

[119] Storck, C. R.; Pereira, J. M., Pereira, G. W.,Rodrigues, A.O., Gularte, M. A., Dias, Á. R. G. (2009). Características tecnológicas de pães elaborados com farinha de arroz e transglutaminase. *Brazilian Journal of Food Technology,* II SSA, 71-77.

[120] Sun-Waterhouse, D., Chen, J., Chuah, C., Wibisono, R., Melton, L. D., Laing, W., Ferguson, L. R., Skinner, M. A. (2009). Kiwifruit-based polyphenols and related antioxidants for functional foods: Kiwifruit extract-enhanced gluten-free bread. *International Journal of Food Sciences and Nutrition, 60, 7,* 251-264.

[121] Tedru, G.A.S., Ormenese, R.C.S.C., Speranza, S.M., Chang, Y.K., & Bustos, F.M. (2001). Estudo da adição de vital glúten à farinha de arroz, farinha de aveia e amido de trigo na qualidade de pães. *Ciência e Tecnologia de Alimentos, 21,1,* 20-25.

[122] Torbica, A., Hadnadev, M., & Dapčević, T. (2010). Rheological, textural and sensory properties of gluten-free bread formulations based on rice and buckwheat flour. *Food Hydrocolloids, 24, 6-7,*626-632.

[123] Toufeili, I., Dagher, S., Shadarevian, S., Noureddinei, A., Sarakbi, M., & Farran, M. T. (1994). Formulation of Gluten-Free Pocket-Type Flat Breads: Optimization of Methylcellulose, Gum Arabic, and Egg Albumen Levels by Response Surface Methodology. *Cereal Chemistry, 71, 6,* 594-601

[124] Vemulapalli, V., Miller, K.A., & Hoseney, R.C. (1998).Glucose oxidase in breadmaking systems. *Cereal Chemistry, 75, 4,* 439-442.

[125] Witczak, M., Korus, J., Ziobro, R., & Juszczak, L. (2010). The effects of maltodextrins on gluten-free dough and quality of bread. *Journal of Food Engineering, 96, 2,* 258-265.

[126] Ylimaki, G., Hawarysh, Z. J., Hardin, R. J., & Thomson, A. B. R. (1988). Application of response surface methodology to the development of rice flour yeast breads: objective measurements. *Journal of Food Science, 53, 6,* 1800-1805.

[127] Young, L., & Cauvain, S. P. (2006). *Baked Products: Science, Technology and Practice.* Ed. Blackwell Publishing Limited.

[128] Young, L., & Cauvain, S. P. (2007). *Technology of Breadmaking.* Ed. Springer; 2nd ed..

[129] Yu, E.Q. (2008). *Starch-free flour for noodles, bread and the like.* Patent References 3348951.

[130] Zandonadi, R. P., Botelho, R. B. A., & Araújo, W. M. C. (2009). Psyllium as a Substitute for Gluten in Bread. *Journal of the American Dietetic Association, 109, 10,* 1781-1784.

[131] Zhang, H.X., & Lugar, P. (2010). *Method and Formulations for Gluten-Free Bakery Products.* IPC8 Class: AA21D1306FI, USPC Class: 426 19, Patent application number: 20100015279.

Chapter 2

IMPACT OF NITROGEN AND SULFUR FERTILIZATION ON GLUTEN COMPOSITION, AND BAKING QUALITY OF WHEAT

Dorothee Steinfurth, Karl H. Mühling and Christian Zörb[*]

Christian Albrechts University Kiel,
Institute of Plant Nutrition and Soil Science, Kiel, Germany

ABSTRACT

Wheat has a broad genetic potential for the expression of various gluten proteins and is strongly influenced by the environment, e. g. with respect to the fertilization of nitrogen (N) and sulfur (S) during wheat plant development. Nitrogen and S availability markedly change wheat flour and baking quality. However, without adequate management of N and S fertilization, the genetic potential of wheat cannot be exploited. Nitrogen fertilization primarily affects the concentration of gliadines and glutenines as it is source-regulated. Storage protein concentration, particularly of gliadine and glutenine, is highly correlated with loaf volume. The use of three instead of one dressing of fertilizer, especially as a late dressing at ear emergence, the gliadine and glutenine concentration of the grain and, furthermore, loaf volume can be enhanced. However, not only quantity, but also the quality of gluten proteins is important for their composition and characteristics. Despite protein amount being only slightly influenced by increasing S fertilization levels, the composition of gluten proteins is highly affected. Increasing S availability for plants enhances S-containing amino acids, namely cysteine and methionine, which are foremost in S-rich gluten proteins. For an improvement of baking quality, a high concentration of S-rich proteins is of particular importance in order to form an appropriate gluten network within the dough. Additionally, a late S dressing can further improve baking quality, since such a dressing increases the synthesis of high molecular weight (HMW) glutenins that considerably affect baking quality in wheat.

S-rich components such as sulfate and glutathione are, moreover, important molecules for the transport and supply of S to the grain. Both these molecules represent important S transport forms from source to sink organs. The potential to synthesize large

---

[*] Corresponding author: Christian Zörb czoerb@plantnutrition.uni-kiel.de

amounts of storage protein depends on the ability of the plant to take up and to transport S into sink organs such as ears and kernels. Sulfur and glutathione are therefore important molecules functioning as actuators during plant growth and grain development. Furthermore, glutathione can function as internal plant signal for the S status of the plant. In terms of baking quality, glutathione is able to interact with gluten proteins responding in a change of existing disulfide bonds between gluten proteins and resulting in a different rheological property and baking quality. Concerning the interaction of N and S in wheat, a well-balanced N/S ratio is exceedingly important for suitable gluten and baking quality. If high N fertilization is accompanied by inadequate S fertilization, S deficiency is provoked, resulting in changed gluten protein composition and a loss of nutritional quality of the grains.

INTRODUCTION

Gluten proteins in cereals belong to the family of plant storage proteins and primarily provide nitrogen (N) and sulfur (S) stores for the growth of seedlings. In terms of human and animal nutrition, these storage proteins are not only a source of essential amino acids, but also of particular importance for product quality. Especially in wheat, gluten proteins are highly important, since they have special characteristics, appearing exclusively in wheat, for dough preparation and baking quality (Zhao et al., 1999b).

Fertilization of wheat plants with N and especially with S has been extensively discussed, since environmental S depositions from the atmosphere have decreased during the last decades (Vestreng et al., 2007). Sulfur dioxide (SO_2) from the atmosphere thus no longer acts as a sufficient source for S. Hence, S application to the soil in any form and its management during plant development needs to be further investigated. Although wheat is not such a high-S-demanding crop plant as canola, S deficiency becomes obvious as soon as S fertilization is inadequate. Sulfur deficieny in wheat results in a change of color from dark green to light-green in young leaves (Bell et al., 1995), whereas N deficiency is diagnosed by the lightening of old leaves. Consequently, an inadequate nutrient supply results in reduced yield (Zhao et al., 1999a), the lower production of S-containing amino acids and S-rich storage proteins (Wieser et al., 2004), and low baking quality. These aspects, when taken together, lead to lower profits for farmers and the baking industry.

Furthermore, as a reinforcing situation, wheat breeding during the last few decades has focused mainly on increasing yield and resistance against plant pathogens and only to a lesser extent on baking quality. As a consequence, yield has increased at the expense of protein concentration in grain (Feil, 1997; Statistisches Bundesamt, 2010).

Each wheat cultivar has its characteristic, genetically determined pattern of gluten proteins, which have a major influence on baking quality. Irrespective of the genetic background, N and S fertilization further highly influence the composition of storage proteins. Thus, fertilization is a means to influence not only grain yield, but also grain quality. As early as the middle of the last century, Finney and Barmore (1948) showed that wheat cultivars exhibit different slopes with respect to loaf volume, and thus baking quality increases with protein concentration independently of the genetic background of the cultivars. Is it therefore possible to increase both yield and baking quality by adequate N and S fertilizer application?

The main focus of this chapter is to demonstrate the impact of N and S fertilization on the modification of gluten proteins and their rheological characteristics and baking quality parameters. Nitrogen and S fertilization can be managed in several ways: (1) the amount of N and S fertilizer applied, (2) the application of soil or foliar fertilizers, and (3) the application at different developmental stages of plant growth. Moreover, an elucidation of the way that N and S interact during the formation of gluten proteins is of interest.

GLUTEN PROTEIN COMPOSITION AND SYNTHESIS

Since T. J. Osborne's (1924) publication concerning the extractability of storage proteins, they have been classified into four groups in wheat: the albumins and globulins, which are mainly metabolic and structural proteins (Goesaert et al., 2005), are extractable with water and NaCl solution, respectively. Other storage proteins such as the prolamin or gluten proteins are divided into gliadins, which are extractable with ethanol, and glutenins, which need to be extracted with reducing agents because of covalent cross-links such as disulfide bonds (Osborne, 1924; Shewry, 1995). Gliadins and glutenins account for about 75 % of mature wheat grain proteins (Shewry and Halford, 2002) and are the most important factors for the remarkable breadmaking properties of wheat.

Since the role of globulins in baking quality is still not fully understood (Shewry and Halford, 2002), this chapter will focus on gliadin and glutenin proteins. Gluten proteins further consist of subunits that contain no (S-poor), low (high molecular weight glutenin subunits, HMW-GS), or high (S-rich) S amounts in the form of cysteine (Shewry and Halford, 2002). Gliadin proteins are monomeric structures that can only form intra-molecular subunits, whereas glutenins are able to form both intra- and inter-molecular subunits. B- and C-type low molecular weight (LMW) subunits of glutenin, and γ- and α-gliadins are S-rich molecules and contain about 2-3% cysteine. D-type LMW subunits of glutenins and ω-gliadins are S-poor and contain only 0 to <0.5% cysteine. Cysteine possesses a free thiol-group that is able to form disulfid cross-links with other cysteine-containing molecules. HMW glutenins contain repetitive domains that may be intrinsically elastic because of ß-turns in the protein structure that are bordered by α-helices on the N- and C-terminal domain of the molecule. Additionally, HMW-GS are notably divided into two types according to their genetic background: y-type HMW-GS that are of low molecular weight and x-type HMW-GS subunits that are of high molecular weight. The y-type HMW-GS contain more SH-groups at the N-terminal domain than the x-type HMW-GS (Shewry et al., 1992). Further information concerning glutenin composition can be obtained by reference to the above-mentioned reviews on HMW glutenin subunits by Shewry et al. (1992) and Shewry and Halford (2002).

Gluten protein expression is highly tissue-specific and does exclusively occur in seeds during grain development. The expression of gluten proteins is further regulated by available N and S metabolites that are synthesized during plant growth (Tabe et al., 2002). Nitrogen is on first instance important for amino acid and protein synthesis. Therefore, the next chapter will discuss the role of N fertilization on gluten composition and its effect on baking qualiy.

The Role of Nitrogen Fertilization in Gluten Formation and the Resulting Baking Quality

Assimilation and Translocation of Nitrogen Within the Wheat Plant

In wheat plant roots, nitrate transporters are mainly responsible for the uptake of nitrate (NO_3) into the root symplast (Yin et al., 2007). Nitrogen is also taken up by wheat roots in the form of ammonium (NH_4^+) (Cox and Reisenauer, 1973). Ammonium, however, cannot be stored in vacuoles in the same manner as nitrate, and furthermore, ammonia (NH_3) is a cell-toxic compound that needs to be converted into amino acids or amides as rapidly as possible. Most plant species prefer to take up N in the form of nitrate, although some species do indeed take up N in the form of ammonium. This can further be observed under special soil conditions, such as low soil pH and low redox potential (Ismunadji and Dijkshoorn, 1971). For an adequate growth rate and yield of wheat, both forms of N (nitrate and ammonium) have to be applied, since the application of only one N form has adverse effects on the uptake of other nutrients and further on pH regulation (van Beusichem et al., 1988).

In brief, N is assimilated as follows. NO_3 is taken up into the cytosol of root cells and further converted to nitrite by nitrate reductase. In the next step, nitrite is transported to chloroplasts or plastids and is converted to NH_4 by nitrite reductase. NH_4 is the cation that is used to form the first amino acids in plant metabolism: glutamine and glutamate. In the plastids (roots) and chloroplasts (leaves), glutamine synthetase and glutamate synthase (GOGAT) convert NH_4 into glutamine and glutamate (Miflin and Habash, 2002). These amino acids are now available to synthesize all the other amino acids and to build up metabolic and structural proteins that are needed during the vegetative growth of wheat plants. Nitrogen assimilation takes place in both root and leaf cells. NO_3 is transported from roots to leaves via the xylem. If NO_3 is present to excess in leaves, it is stored in vacuoles and can be regenerated as soon as NO_3 concentrations decrease in the cytosol. During generative growth, NO_3 concentrations in the soil without additional fertilization is not sufficiently supplied, and the high demands of sink tissues cannot be met. Therefore, plant organs, especially the flag leaf, regenerate N from protein degradation to meet the N demands of the developing grain (Gooding et al., 2007). Mainly glutamine and asparagine are transported to the grains via the phloem to synthesize storage proteins (Macnicol, 1977; Anderson and Fitzgerald, 2001). The wheat grain has different potentialities for attaining N. First, as discussed above, N that has been stored during vegetative growth can be mobilized, or second, further N can be taken up directly from the soil (deRuiter and Brooking, 1996). If N is not limiting, the primary source is N from vegetative tissues (60-92%) that has previously been assimilated (Simpson et al., 1983). Furthermore, Martre et al. (2003) have found N in grain to be source-regulated, i. e. total grain N is regulated by N in vegetative tissues as incorporated prior to anthesis (Barneix and Guitman, 1993; Fuertes-Mendizábal et al., 2010) but not by the activity of the grain namely the sink tissue. However, increasing N fertilization enhances the N concentration in the leaves of wheat and further provides the potential to obtain higher amounts of N for translocation to the grain and therefore the synthesis of gluten proteins.

Nitrogen Fertilization Increases the Amount of Gluten Proteins. is This All We Know?

Nitrogen fertilization increases the protein concentration in all wheat plant tissues, and especially in wheat grain (Finney and Barmore, 1948). First and foremost, an increase in N fertilization enhances grain yield. However, Gauer et al. (1992) and Garrido-Lestache et al. (2005) have observed stagnating grain yield after N rates of 120 kg ha^{-1} and 100 kg ha^{-1}, respectively, despite protein concentration still increasing. The enhancement of grain protein concentration by incremental additions of N fertilizer is limited in the same way (Barneix et al., 1992; Triboi and Triboi-Blondel, 2002). After reaching a maximum N concentration in grain, only the N concentrations within the vegetative plant parts increase on additional N application. Mechanisms mediated by ammonium and glutamine reduce the expression of nitrate transporters in roots by negative feedback reactions. As long as storage protein synthesis is in progress, ammonium and glutamine concentrations will remain low and result in a constant N uptake (Foulkes et al., 2009). Furthermore, application at different stages of plant development influence grain yield and N concentration. Early N fertilization during vegetative growth enhances grain yield by influencing kernels per ear and ears per plant. Nitrogen fertilization at anthesis predominantly benefits grain N and therefore grain protein concentration (Jahn-Deesbach and Weipert, 1964; Wuest and Cassman, 1992).

Investigations of gluten protein fractions have shown that Osborne fractions are differently affected by N fertilization. Unlike gliadin and glutenin proteins, albumins and globulins are hardly influenced (Wieser and Seilmeier, 1998; Fuertes-Mendizábal et al., 2010). Wieser and Seilmeier (1998) have provided further insight into the gluten protein composition of various wheat cultivars that were raised on different N regimens. Interestingly, the quantity of gliadins increases more than the quantity of glutenins. These findings have been confirmed by Triboi et al. (2000) in field experiments.

With respect to the pattern of gliadin and glutenin subunits, the quantity of S-poor gluten proteins such as ω-gliadins and HMW-glutenins is mainly enhanced after increasing N fertilization. Wieser and Seilmeier (1998) have further investigated the HMW subunits, and observed that x-type glutenins are more influenced by N fertilization than y-type glutenins. X-type glutenins contain fewer thiol-groups than do y-type glutenins (Shewry et al., 1992). As a consequence, N fertilization enhances primarily S-poor subunits by increasing the expression of S-poor gluten subunit genes. Prieto et al. (1992) have observed increasing ω-gliadin concentrations with increasing N fertilization in agreement with the findings by Wieser and Seilmeier (1998).

Recent investigations concerning late N fertilization have provided new insights concerning gluten protein modification. Late fertilization has the following advantages: late N fertilization at ear emergence enhances protein concentration and, furthermore, recovers deficiency symptoms, since fast nutrient translocation mechanisms exist in plants. For example, Foulkes et al. (2009) have observed that N deficiency symptoms at anthesis can be even recovered by a later N fertilization. Tea et al. (2007) have examined the translocation of N and S after a late foliar application and found high recovery rates in leaves and grain with a beneficial impact on N/S ratios and protein concentration. Therefore, under inconvenient weather conditions with high or no precipitation, a split application of nutrients prevents leaching or helps to enable nutrient uptake by foliar application, respectively. Furthermore, the application of N at a time point when demand for N is high makes sense. This high demand occurs predominantly during the first few days of grain development, when proteins have started to be synthesized (Emes et al., 2003; Shewry et al., 2009).

Unlike increasing N fertilization treatments that predominantly increase gliadin concentration, a late application of N, when the flag leaf is just visible, mainly influences the concentration of glutenins (Fuertes-Mendizábal et al., 2010). With respect to the subunits of gluten proteins, each HMW-GS evidently increases with increasing N fertilization. Although HMW glutenins are minor components of gluten proteins, compromising approximately 12% of all gluten proteins, these glutenin subunits notably account for 45-70% of variation in baking quality (Shewry et al., 2000). This aspect largely explains the variation in baking quality after different N fertilization treatments.

Nitrogen Fertilization Increases Baking Quality, But the Question is How?

As early as 60 years ago, Finney and Barmore (1948) showed convincingly that the protein concentration in wheat flour of individual cultivars is highly correlated with loaf volume. However, the parameter that is responsible for this high correlation remains unknown. Parameters such as the gliadin/glutenin ratio (MacRitchie, 1980; Reinbold et al., 2008), amino acid composition (Granvogl et al., 2007), and thiol and disulfide status (Graveland et al., 1978) have been considered as possible explanations for the correlation but have failed to account for the correlation completely. The interdependency seems to be more complex than can be explained by only one parameter. However, can single gluten subunits indeed explain the impact on baking quality?

Many direct and indirect methods are available to examine baking quality on dough or on gluten proteins. The standardized baking test, for example the rapid-mix-test (RMT) (Pelshenke et al., 1978) is a direct method for examining loaf volume, dough consistency during preparation, and sensory parameters. Other methods such as sedimentation value, extensigram, alveogram, and mixograph-test are only capable of predicting loaf volume by the correlation of the flocculation of gliadins and glutenins, elasticity, resistance, and mixing properties, respectively, with loaf volume. Despite the standardized baking test being the only direct method for obtaining information about loaf volume, particularly in small-scale investigations such as pot experiments, indirect methods are appropriate for providing information about the impact of single gluten proteins on rheological parameters.

Several investigations have examined both the impact of N fertilization on baking quality parameters and the change of individual gluten proteins (Luo et al., 2000; Flaete et al., 2005; Fuertes-Mendizábal et al., 2010). Flæte et al. (2005) have observed increasing sedimentation values and resistance of dough (R_{max}) in gluten extensigraph experiments when comparing high N with low N fertilization levels. A diminished S concentration in gluten proteins as a result of inadequate S fertilization has been assumed to explain the increased resistance of dough. To underline the obvious impact of S, Popineau et al. (1994) used balanced N and S levels and concluded that this combination did not result in increased R_{max}.

Furthermore, two-dimensional (2D)-gel electrophoresis has been used to explain the effect of increased N fertilization on protein composition. Flæte et al. (2005) have found two proteins to be changed by increased N fertilization. However, these proteins have been identified to be non-gluten proteins such as glyceraldehyde-3-phosphate dehydrogenase and one serpin protein. With respect to the sedimentation value, Luo et al. (2000) have also compared early N fertilization at booting with late N fertilization at flowering and have observed the same positive impact of N fertilization on the sedimentation value as that reported by Flæte et al. (2005), irrespective of whether either early or late N application was conducted. Unlike Flæte et al. (2005), who used extensibility tests on dough to elucidate the impact of single gluten proteins on baking properties, Luo et al. (2000) used a mixograph test. The mixograph test revealed that the mid-line peak value, but not the mid-line peak time, was enhanced with N fertilization. To conclude, both groups of authors found an increase of dough resistance and a decrease of extensibility attributable to N fertilization. These quality parameters indicate reduced baking quality, since higher resistance and lower extensiblity account for a dough that is firm and therefore proves less well on dough fermentation.

Are there any hints in gluten composition that can explain the increased resistance? Ayoub et al. (1994), for example, have investigated both N rate and timing on loaf volume and found that an increasing N rate and late N fertilization increases both, the protein concentration and the loaf volume. Evidently, increasing N fertilization produces wheat doughs with lower extensibility and higher resistance, since those gluten proteins, which are poor in S, increase. Fuertes-Mendizábal et al. (2010) have further elucidated the effect of a late N application on baking properties and additionally found explanations in the gluten composition. Baking quality parameters examined on an alveograph increase with both an increasing N rate and a late N application. Dough strength increases by 50% with increasing N rate, whereas splitting enhances dough strength by about 80%. A late N application obviously has pivotal effects on both total protein concentration and baking quality characteristics. Investigations into the effects of late N fertilization on gluten composition by Fuertes-Mendizábal et al. (2010) have further enlightened the impact of individual proteins on dough characteristics. The x-type HMW-GS, which contain a smaller amount of SH-groups, are more affected by N fertilization than are y-type HMW-GS. Furthermore, late N fertilization seems to enhance all individual HMW-GS and thus increases dough strength and extensibility (Fuertes-Mendizábal et al., 2010). However, an evaluation of the effect of N fertilization on baking quality parameters remains difficult, since the S effect cannot be excluded from the evaluation.

INFLUENCE OF S FERTILIZATION ON GLUTEN COMPOSITION AND BAKING QUALITY

Assimilation and Translocation of Sulfur Within the Wheat Plant

Since S depositions from the atmosphere have decreased over the last few decades, the most important source of S for plants is the sulfate in soil, which is taken up by roots via sulfate transporters (Clarkson et al., 1993; Hawkesford et al., 1993; Buchner et al., 2004). Instead of assimilation in root cells, sulfate is, to a higher extent, transported to leaves for storage or direct assimilation (Larsson et al., 1991). Comparable to N assimilation, sulfate is stored in vacuoles as soon as sulfate uptake exceeds the rates of S assimilation. The first step of S assimilation is the activation of sulfate by ATP resulting in the formation of adenosine phosphosulfate. In the various subsequent steps, adenosine phosphosulfate is bound to thiol groups and is then reduced to sulfite (SO_3^{2-}) and further to sulfide (S^{2-}). All these assimilation steps are located in the chloroplasts of wheat leaf cells. In the next step, H_2S is incorporated into O-acetyle-serine to form the first S-containing amino acid, cysteine (Schiff and Saidha, 1987). Cysteine is further transported from chloroplasts to the cytosol to form methionine or is combined with glycine and glutamine to form the S-containing tripeptide glutathione (Ravanel et al., 1998; Hesse and Hoefgen, 2003). Glutathione fulfills several functions in plant metabolism, such as the reduction of oxidative stress within the glutathione-ascorbate cycle (Noctor and Foyer, 1998) and the detoxification of xenobiotics (Schröder et al., 2007) and heavy metals (Cobbett and Goldsbrough, 2002). Additional functions concern S translocation in wheat plants. Glutathione is one component that is transported from vegetative tissues, predominantly the flag leaf, to the developing grain (Rennenberg et al., 1979). However, the component that is primarily transported to the grain is not yet completely known. The S-containing compounds sulfate, glutathione, and S-methyl-methionine have been found in phloem sap during grain development (Bonas et al., 1982; Larsson et al., 1991). However, in wheat, S-methyl-methionine is presumed to be the main transport form of S to developing grains (Bourgis et al., 1999). Grains seem to prefer S-containing metabolites for the formation of S-rich gluten proteins (Steinfurth et al., submitted). Another option involves the direct assimilation of sulfate that is taken up from the soil and then directly transported to the grain (Steinfurth et al., submitted). Nevertheless, grains are able to take up sulfate directly from the xylem stream which has been revealed by investigations on sulfate transporter expression in wheat plant tissues (Buchner et al., 2004). Recently, Buchner et al. (2010) have further elucidated the impact of vegetative plant tissues on the translocation of S within the plant during grain development. If the duration of grain development and storage protein synthesis is prolonged, an increasing number of leaves is responsible for maintaining S transport toward the grain. Young leaves and primarily the flag leaf are not able to meet the S demand of the grain, and therefore, older leaves have to mobilize S to supply the sink organ, *viz.*, the grain.

Furthermore, the uptake and translocation of S is highly regulated. Endless absorption of S by the roots during vegetative growth is impossible, since sulfate transporters undergo a negative feedback mechanism (Saito, 2000). Nevertheless, the compound in plant metabolism that is responsible for the negative feedback on sulfate transporters is not yet known. Anderson and McMahon (2001) have assumed sulfate or glutathione to down-regulate the transporter expression. Additional investigations have revealed that glutathione alone is not able to repress the transporter expression. Thus, the ratio of sulfate to glutathione is suggested to act more probably as an internal signal (Rennenberg, 2001).

Grain usually only contains small amounts of sulfate (Steinfurth et al. submitted). Unlike vegetative tissues that contain about 50% of total S as sulfate, grains only contain about 1-5% of the total S as sulfate (Zhao et al., 1999a). Nevertheless, the way that wheat grain forms amino acids and polypeptides with a special function in bread making is of interest.

Does Sulfur Fertilization Increase Gluten Proteins as it Has Been Reported for Nitrogen Fertilization?

Increasing S fertilization is accompanied by a higher S concentration in grain (Zhao et al., 1999a; Zörb et al., 2009). High S amounts are incorporated into the S-containing amino acids cysteine and methionine, which accumulate in grain during development (Zhao et al., 1999a).

Unlike replete S-supplied plants that predominantly store glutamate as a main amino acid in grain, S-starved plants accumulate the amino acid asparagine. This amino acid has devastating effects on the nutritional quality of bakery products. As the authors Granvogl et al. (2007) have shown, acrylamide formation increases during the baking process since concentrations of free asparagine and sugars are enhanced because of S deficiency.

However, next to the negative effects of product processing, the protein quality of wheat is also highly important for human and animal nutrition since cereal products cover major amounts of daily protein intake (Max Rubner-Institute, 2008). To examine the biological value of each essential amino acid in wheat flour the protein digestibility corrected amino acid score (PDCAAS) was calculated (Schaafsma, 2000). For this purpose a pot experiment with increasing S fertilization (0 g S, 0.1 g S, and 0.2 g S at sowing) and, in particular, a late S fertilization (0.1 g S at sowing and additionally 0.1 g S at ear emergence) was conducted. This score has the advantage of evaluating the biological quality of protein by considering the characteristic digestibility and requirements for a 5-year-old child. The PDCAAS was not affected by S fertilization (Fig. 1). Unlike S fertilization, N fertilization decreases lysine concentrations (Shewry, 2007) because higher gluten concentrations are accompanied with a low lysine abundance (Kasarda et al., 1978). Nevertheless, the PDCAAS of cysteine and methionine increases with S fertilization (Fig. 1). Methionine is, in contrast to lysine, not a limiting amino acid in wheat but is essential for human and animal nutrition. The facilitation of an adequately S-fertilized wheat might represent a chance to enhance the amount of methionine in the human diet and, hence, the protein quality consumed. Since humans and animals consume not only wheat-based products, high methionine levels in wheat can compensate for other low methionine sources.

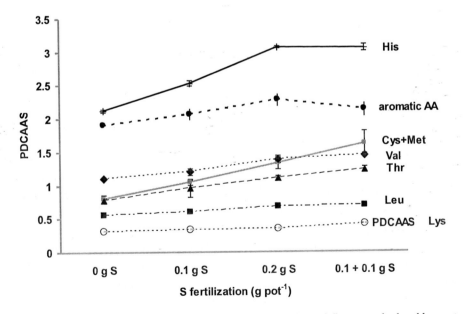

Figure 1. Impact of S fertilization on the biological value of wholemeal flour as calculated by protein digestibility corrected amino acid score (PDCAAS). Plant cultivation of cultivar Türkis was conducted according to Zörb et al. (2009). The S fertilization of the cultivar is indicated on the x-axis including a late S dressing (0.1 + 0.1 g S). PDCAAS was calculated according to Schaafsma (2000). His: histidine, aromat. AA: aromatic amino acids Tyr+Phe, Cys+Met: cysteine+methionine, Val: valine, Thr: threonine, Leu: leucine, Lys: lysine.

Unfortunately, a further negative parameter concerns the correlation of the lysine concentration in wheat protein with baking quality. Nitrogen fertilization predominantly enhances the gluten protein concentration (Wieser and Seilmeier, 1998; Fuertes-Mendizábal et al., 2010), but does not affect the albumin and globuline fractions. Since these fractions contain higher lysine amounts than gluten fractions (Shewry, 2009), increasing baking quality is negatively correlated with nutritional protein quality. However, the question arises as to whether S fertilization changes gluten composition, and whether these changes influence baking quality.

In freshly prepared flour, 95% of cysteine is present as inter- or intramolecular disulfides and only 5% as the thiol form (Shewry and Tatham, 1997; Grosch and Wieser, 1999). These disulfide bonds are predominantly formed between LMW and HMW glutenin subunits resulting in gluten-protein aggregates. Therefore, S fertilization changes the protein composition and baking quality mainly by increasing the cysteine concentration in gluten proteins. Unlike N fertilization that predominantly affects the quantity of protein in grain, S fertilization is known to "fine-tune" gluten proteins (Wieser et al., 2004; Zörb et al., 2009).

With respect to the Osborne fractions, the quantity of total gliadin and glutenin proteins is not affected by increasing S fertilization (Zörb et al., 2010). Unlike plants that receive no or little S, well-fertilized plants synthesize large amounts of S-rich γ-gliadin and LMW-glutenin proteins (Wieser et al., 2004; Zörb et al., 2009). Accordingly, S-poor gluten proteins such as ω-gliadins and HMW-glutenins are reduced. However, how can the reduced synthesis of S-rich gluten proteins be explained by plant metabolism?

An adequate S supply during wheat plant development enhances the S pool in leaves and provides large amounts of S available for translocation to the grain. Furthermore, large amounts of glutathione are synthesized in the leaves (Steinfurth et al., submitted), and proteins might contain higher amounts of cysteine and methionine. Therefore, high concentrations of glutathione (Rennenberg, 1982) and S-methylmethionine (Bourgis et al., 1999) are translocated to the grain via the phloem. Exclusively under S deficiency, protein degradation in leaves is necessary to meet the high S sink capacity of grains (Fitzgerald et al., 1999). Because of high amounts of glutathione available in grain, S-rich proteins are primarily synthesized (Shewry et al., 2001). As long as large amounts of S are supplied to the grain, S-rich gluten proteins such as LMW glutenin subunits and γ-gliadins are predominantly synthesized (Wieser et al., 2004; Zörb et al., 2009). The higher the S concentration of the grain, the higher the LMW glutenin concentration, and the lower the HMW glutenin concentration (Wieser et al., 2004). As a result, the HMW/LMW ratio increases. HMW glutenin subunits are thought to interact with other HMW or LMW subunits to form glutenin aggregates (Shewry et al., 1992). Several investigations concerning polymer formation have concluded that the associations between glutenin subunits are non-random. Glutenin subunits encoded by chromosome 1D are extensively involved in oligomer formation, and variants of genome 1B subunits differ in their ability to combine with 1A and 1D subunits (Lawrence and Payne, 1983). Furthermore, x-type subunits are able to form inter-chain bonds with two cysteine molecules, whereas y-type subunits use six thiol-groups for the formation of aggregates (Moonen et al., 1985). However, these aggregation models are still under discussion and need to be examined further in detail.

Higher S fertilization evidently leads to greater amounts of S being translocated to the grain, and therefore, a higher potential for aggregates in the dough. Without the aggregation of gluten proteins, wheat dough has no ability to gain volume during dough fermentation, since the matrix formation containing predominantly starch and gluten proteins is restricted (Gan et al., 1995). However, how does a late S fertilization or foliar application of S influence gluten composition? The impact of a late S fertilization on gluten composition has recently been investigated. The effects of late S fertilization on baking quality characteristics are highly controversial. Zörb et al. (2009) have observed no effects of late S fertilization on the protein and S concentration in wheat grain compared with grain obtained following fertilization with the same amount of S applied at sowing. Even more controversially, Luo et al. (2000) have concluded that a late S application is not necessary to optimize baking quality, but glutenin proteins respond to both late S and combined late N and S fertilization with quality reduction.

The experiment of Luo et al. (2000) has nevertheless a shortcoming concerning the N and S application time. Only the impact of N and S fertilization and its combination either at booting or at flowering has been examined. Hence, wheat plants might be grown on N- and S-deficient soil until booting and flowering, which has an important influence on plant metabolism in vegetative and generative tissues. To prevent quality losses, Haneklaus et al. (1995) concluded from S-starvation studies that S has to be supplied before S deficiency symptoms become apparent. Furthermore, Zörb et al. (2009) have demonstrated that a late S fertilization at ear emergence prevents S deficiency attributable to low S application at sowing.

By comparing S fertilization at sowing with the split application of S, *viz.*, at sowing and at ear emergence, a late S application effectively results in an increased loaf volume (Seling et al., 2006; Zörb et al., 2009). Reversed phase-high performance liquid chromatography (RP-HPLC) and the 2D-gel electrophoresis of flour proteins might provide further insight into the alteration of gluten protein composition after late S fertilization. Zörb et al. (2009) have found HMW glutenin subunits of type 1By9 and 1Bx7 to be enhanced after late S fertilization. However, Grove et al. (2009) have identified LMW glutenins and γ-gliadins as being enhanced after late S fertilization at early heading. Unfortunately, glutenins of high molecular weight were excluded from their analysis, because HMWs were negatively stained.

In addition, late S fertilization changes the protein profile of wheat grain during whole grain development time. Figure 2 illustrates the changes in protein pattern of wheat grain with respect to milk ripe kernels (13 days post anthesis) and mature wheat grain. Grain material of milk ripe and mature kernels from a pot experiment with the winter wheat cultivar Türkis was harvested, and 2D-gel overlays of high S fertilization (0.2 g S at sowing) with late S fertilization (0.1 g S at sowing and 0.1 g S at ear emergence) were compared (Fig. 2). However, notably the protein profile predominantly changes in mature but not in milk ripe grain. In mature grain, more proteins are up-regulated in comparison with those in milk ripe grain. To conclude, S fertilization has an severe impact on the protein profile during the whole grain development and appears to change protein synthesis mainly during later stages of grain development. Shewry et al. (2009) have further illustrated protein synthesis during grain development by using 1D-gels and have demonstrated that LMW and HMW-GS are synthesized continuously until maturity, whereas other gluten proteins, such as ω-gliadins do not eventually attain further concentration. Since protein synthesis occurs until grain maturity and continues during the dessication phase, changes in S fertilization appear to regulate gluten protein gene expression until maturity.

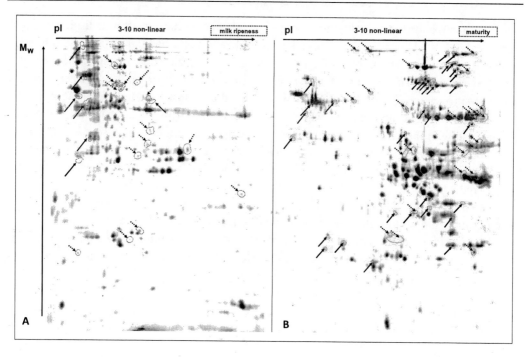

Figure 2. Impact of S fertilization on protein synthesis in various developmental stages of wheat grain. Overlays of wheat flour proteins comparing high with late S fertilization. A) Overlay of images of two-dimensional (2D)-gel electrophoresis of immature milk ripe grain and B) overlay of flour of mature wheat grain. Continuous (rising) arrows illustrate the higher quantity of protein with fertilization levels after late S application, dotted (falling) arrows indicate higher protein quantity in plants with 0.2 g S at sowing. Wheat plants were grown under high (0.2 g S at sowing) and late S fertilization (0.1 g S at sowing + 0.1 g S at ear emergence). Kernels were harvested at milk ripeness and maturity. Pot experiment with cultivar Türkis and overlays of 2D gels comparing high S with late S fertilization were conducted according to Zörb et al. (2009).

To date, further investigations, especially those under field conditions with diverse fertilization techniques involving soil or foliar application are necessary to elucidate completely the impact of a late S fertilization on gluten composition and baking quality characteristics. Foliar application of S, for example, is a strategy to prevent S deficiency in the late growth stages of wheat, since foliar applied S is still incorporated in grain gluten proteins (Tea et al., 2003). In addition to the impact of foliar S application on S translocation, Tea et al. (2005) have elucidated the impact of foliar S application on the thiol-disulfide status and polymeric protein formation in wheat grain. Glutathione is an interacting molecule in gluten formation during grain development (Hüttner and Wieser, 2001; Tea et al., 2005). High levels of reduced glutathione (GSH) decrease the maximum resistance of dough and concomitantly increase extensibility (Kieffer et al., 1990). Glutathione subsequently interacts with thiol-groups of gluten proteins and changes their properties. In particular, during the desiccation phase of grain development, the content of polymeric proteins increases. Furthermore, the impact of S fertilization on glutathione concentration and the polymerization of gluten proteins becomes noticeable during grain development (Tea et al., 2005). With low S fertilization, low polymeric protein-bound-glutathione is available and further low aggregated proteins emerge. Glutathione binds exclusively to cysteine residues that form inter-molecular disulfide bonds and are consequently responsible for the formation of LMW-

GS aggregates. However, the effect of glutathione on gluten composition and its results in baking quality are not yet completely understood. Further investigations are needed. Reduced glutathione is known to affect the rheological properties of dough by the depolymerization of glutenin polymers via the specific cleavage of inter-molecular disulfide bonds. Reinbold et al. (2008) have elucidated the effect of increasing S fertilization on the thiol and glutathione status of S-deficient wheat. Since the dough made from S-deficient flour is highly resistant to expansion, it is of interest to determine the way that S fertilization changes SH-groups in wheat flour. With increasing S fertilization, cysteine and glutathione levels are enhanced. These authors have revealed that, in S-deficient wheat cysteine and glutathione correlated with bread volume in the same extent as the protein concentration. So far, S fertilization is known not only to enhance glutathione concentration in wheat flour, but concomitantly also the amount of S-containing gluten proteins. There is a need to elucidate further whether the impact of glutathione on gluten proteins is different between low S plants and high S plants.

Is There a Potential to Increase Baking Quality Further by Adequate Sulfur Fertilization?

Whereas an examination of the individual impact of N fertilization on baking quality parameters is difficult, the impact of S fertilization on baking quality has been demonstrated in several investigations. So far, S concentration in grain alone is not able to explain all the impact on loaf volume. An increase of the S concentration in grains, accompanied by adequate N fertilization, increases the extensibility of dough while doughs become less short and resistance decreases (Kettlewell et al., 1998; Zhao et al., 1999b; Flæte et al., 2005). Accordingly, S deficiency leads to high resistant doughs with low extensibility (Wrigley et al., 1984). These quality characteristics account for the higher potential of doughs to prove during dough fermentation because of the disulfide bonds in the gluten structure (Shewry et al., 1992), resulting in higher loaf volume (Moss et al., 1981; Flæte et al., 2005) among adequate S fertilization. Zhao et al. (1999b) have shown a positive but low correlation of S concentration in grain, with loaf volume. Therefore, grain S concentration more clearly explains an increasing dough extensibility. Furthermore, the gel-protein also called glutenin macro polymer, which is difficult to extract from wheat flour because of its high aggregation behavior, has a higher correlation with dough extensibility than the S concentration by itself. Haneklaus et al. (1992) have also been able to explain 40% of the variability in loaf volume based on grain S concentration. Reinbold et al. (2008) have further found cysteine and glutathione concentrations in S-deficient wheat to be positively correlated with R_{max}, extensibility, and extension area. Thus, these data confirm the way in which thiol groups and glutathione interact with S-rich gluten proteins and explain the changes in baking properties. As another important aspect, late S fertilization is considered to affect baking quality positively. Seling et al. (2006) have enhanced RMT loaf volume of winter wheat with a late S fertilization by 117 mL 100 g^{-1} flour. The enhanced baking quality may be refered to an increase of HMW subunits 1Bx7 and 1By9 glutenins found to be increased after late S fertilization (Zörb et al., 2009). This combination of proteins was formerly demonstrated to have a strong positive relationship with the Zeleny sedimentation value and dough resistance and therefore with high baking quality (Branlard and Dardevet, 1985; Seilmeier et al., 1991).

Tea et al. (2007) have examined the effect of foliar N and S application as a late application at anthesis after soil N treatment with ammonium nitrate during vegetative growth. Flour protein content was increased each with N and S treatments. However, combined N and S foliar applications resulted in the highest protein concentration. Alveograph measurements confirmed previous studies with soil-applied fertilizers. Nitrogen or combined N and S treatments increased dough strength, whereas S and combined N and S treatments increased dough swelling and extensibility. Hence, foliar N and S application at anthesis is an alternative to soil application during late wheat development, with applied nutrients still being assimilated and incorporated into gluten proteins.

An increase in the quantity of gluten proteins leads to a higher potential of dough to expand during dough fermentation, since more gas cells develop during mixing surrounded by the characteristic starch-protein-matrix (Gan et al., 1995). Furthermore, cysteine residues within this protein-matrix are able to asure extensibility and lower resistance. A balance between extensibility and resistance is of particular importance, since exessive resistance of dough limits expansion, and excessive elasticity leads to a failure of the dough to retain CO_2 during fermentation (Zhao et al., 1997). An adequate relationship of N and S has therefore to be maintained for adequate baking results, and further increases in baking quality are possible by the addition of adequate S fertilization concomitantly with adequate N fertilization.

THE WAY THAT NITROGEN AND SULFUR INTERACT IN GLUTEN COMPOSITION

A range of investigations highlights the importance of the adequate management of concomitant N and S fertilization. For example, Zörb et al. (2010) have found total protein and as well gliadin and glutenin concentrations to increase in flour following the N fertilization of wheat. In addition, a concomitant increase in S fertilization has no effect on total protein concentrations. However, increasing N fertilization without adequate S fertilization can provoke S-deficiency symptoms, despite an increase in yield and protein concentration. Moreover, Eriksen and Mortensen (2002) have examined the effect of S fertilization on barley; S-containing amino acids cysteine and methionine increase following S fertilization, but N concentration decreases because of a dilution effect caused by increased yield. Hence, baking quality can only be enhanced by increasing N fertilization levels providing that the S abundance in proteins is adequate. The question is: what exactly is meant by "adequate"? Randall et al. (1981) and Zhao et al. (1999a) have stated that N/S ratios of 17:1 and higher in wheat grain indicate S deficiency. Loaf volume can be enhanced by N fertilization at first, but as soon as the N/S ratio increases to inadequate levels, S fertilization needs to be adjusted. On the contrary, the S concentration in grain depends on the protein concentration, since the abundance of N and S in gluten protein subunits is genetically manifested. A high S concentration in grain consequently occurs when both high N and high S supply is applied (Moss et al., 1981). Throughout plant development the abundance of metabolites following N and S assimilation determines the expression of individual gluten protein genes (Tabe et al., 2002). Therefore, splitting the rates of N and S or of different fertilizer forms, such as soil or foliar applications, are alternatives to guarantee nutrient availability throughout plant and grain development. Fast mobilization and retranslocation rates within the plant enable the reliable assembly of late-applied N and S into storage proteins and lead to high quality gluten protein properties during breadmaking.

ACKNOWLEDGMENTS

The authors thank Stephanie thor Straten, Fabian Braukmann and Bärbel Biegler for technical assistance. Financial support by the German Research Foundation (DFG Ge 1009/4-1) is gratefully acknowledged.

REFERENCES

Anderson J. W. & Fitzgerald M. A. (2001) Physiological and metabolic origin of sulphur for the synthesis of seed storage proteins. *Journal of Plant Physiology, 158,* 447-456.

Anderson J. W. & McMahon P. J. (2001) The Role of Glutathione in the Uptake and Metabolism of Sulfur and Selenium. In D. Grill, M. Tausz and L. J. de Kok (Eds.), *Significance of Glutathione in Plant Adaptation to the Environment* (pp. 57-99). Dordrecht: Kluwer Academic Publishers.

Ayoub M., Guertin S., Fregeaureid J., Smith D. L. (1994) Nitrogen-fertilizer effect on breadmaking quality of hard red spring wheat in Eastern Canada. *Crop Science, 34,* 1346-1352.

Barneix A. J., Arnozis P. A., Guitman M. R. (1992) The regulation of nitrogen accumulation in the grain of wheat plants (*Triticum aestivum*). *Physiologia Plantarum, 86,* 609-615.

Barneix A. J. & Guitman M. R. (1993) Leaf Regulation of the Nitrogen Concentration in the Grain of Wheat Plants. *Journal of Experimental Botany, 44,* 1607-1612.

Bell C., Fordham M., Richardson P., Cram J., Tomos D. (1995) Cellular and subcellular compartmentation of sulfate in leaves in relation to low-sulfur mobility. *Journal of Plant Nutrition and Soil Science, 158,* 63-65.

Bonas U., Schmitz K., Rennenberg H., Bergmann L. (1982) Phloem Transport of Sulfur in *Ricinus*. *Planta, 155,* 82-88.

Bourgis F., Roje S., Nuccio M. L., Fisher D. B., Tarczynski M. C., Li C. J., Herschbach C., Rennenberg H., Pimenta M. J., Shen T. L., Gage D. A., Hanson A. D. (1999) S-methylmethionine plays a major role in phloem sulfur transport and is synthesized by a novel type of methyltransferase. *Plant Cell, 11,* 1485-1497.

Branlard G. & Dardevet M. (1985) Diversity of grain protein and bread wheat quality. 2. Correlation between high molecular-weight subunits of glutenin and flour quality characteristics. *Journal of Cereal Science, 3,* 345-354.

Buchner P., Parmar S., Kriegel A., Carpentier M., Hawkesford M. J. (2010) The sulfate transporter family in wheat: tissue-specific gene expression in relation to nutrition. *Molecular Plant, 3,* 374-389.

Buchner P., Takahashi H., Hawkesford M. J. (2004) Plant sulphate transporters: co-ordination of uptake, intracellular and long-distance transport. *Journal of Experimental Botany, 55,* 1765-1773.

Clarkson D. T., Hawkesford M. J. and Davidian J.-C. (1993) Membrane and long-distance transport of sulfate. In L. J. de Kok, I. Stulen, H. Rennenberg, C. Brunold and W. E. Rauser (Eds.), *Sulphur Nutrition and Assimilation in Higher Plants* (pp. 3-19). The Hague: SPB Academic Publishing.

Cobbett C. & Goldsbrough P. (2002) Phytochelatins and metallothioneins: roles in heavy metal detoxification and homeostasis. *Annual Review of Plant Biology, 53*, 159-182.

Cox W. J. & Reisenauer H. M. (1973) Growth and ion uptake by wheat supplied nitrogen as nitrate, or ammonium, or both. *Plant and Soil, 38*, 363-380.

deRuiter J. M. & Brooking I. R. (1996) Effect of sowing date and nitrogen on dry matter and nitrogen partitioning in malting barley. *New Zealand Journal of Crop and Horticultural Science, 24*, 65-76.

Emes M. J., Bowsher C. G., Hedley C., Burrell M. M., Scrase-Field E. S. F., Tetlow I. J. (2003) Starch synthesis and carbon partitioning in developing endosperm. *Journal of Experimental Botany, 54*, 569-575.

Eriksen J. & Mortensen J. V. (2002) Effects of timing of sulphur application on yield, S-uptake and quality of barley. *Plant and Soil, 242*, 283-289.

Feil B. (1997) The inverse yield-protein relationship in cereals: possibilities and limitations for genetically improving the grain protein yield. *Trends in Agronomy, 1*, 103-119.

Finney K. F. & Barmore M. A. (1948) Loaf volume and protein content of hard winter and spring wheats. *Cereal Chemistry, 25*, 291-312.

Fitzgerald M. A., Ugalde T. D., Anderson J. W. (1999) S nutrition affects the pools of S available to developing grains of wheat. *Journal of Experimental Botany, 50*, 1587-1592.

Flæte N. E. S., Hollung K., Ruud L., Sogn T., Færgestad E. M., Skarpeid H. J., Magnus E. M., Uhlen A. K. (2005) Combined nitrogen and sulphur fertilisation and its effect on wheat quality and protein composition measured by SE-FPLC and proteomics. *Journal of Cereal Science, 41*, 357-369.

Foulkes M. J., Hawkesford M. J., Barraclough P. B., Holdsworth M. J., Kerr S., Kightley S., Shewry P. R. (2009) Identifying traits to improve the nitrogen economy of wheat: Recent advances and future prospects. *Field Crops Research, 114*, 329-342.

Fuertes-Mendizábal T., Aizpurua A., Gonzalez-Moro M. B., Estavillo J. M. (2010) Improving wheat breadmaking quality by splitting the N fertilizer rate. *European Journal of Agronomy, 33*, 52-61.

Gan Z., Ellis P. R., Schofield J. D. (1995) Gas cell stabilization and gas retention in wheat bread dough. *Journal of Cereal Science, 21*, 215-230.

Garrido-Lestache E., Lopez-Bellido R. J., Lopez-Bellido L. (2005) Durum wheat quality under Mediterranean conditions as affected by N rate, timing and splitting, N form and S fertilization. *European Journal of Agronomy, 23*, 265-278.

Gauer L. E., Grant C. A., Gehl D. T., Bailey L. D. (1992) Effects of nitrogen fertilization on grain protein content, nitrogen uptake, and nitrogen use efficiency of 6 spring wheat (*Triticum aestivum* L) cultivars, in relation to estimated moisture supply. *Canadian Journal of Plant Science, 72*, 235-241.

Goesaert H., Brijs K., Veraverbeke W. S., Courtin C. M., Gebruers K., Delcour J. A. (2005) Wheat flour constituents: how they impact bread quality, and how to impact their functionality. *Trends in Food Science & Technology, 16*, 12-30.

Gooding M. J., Gregory P. J., Ford K. E., Ruske R. E. (2007) Recovery of nitrogen from different sources following applications to winter wheat at and after anthesis. *Field Crops Research, 100*, 143-154.

Granvogl M., Wieser H., Koehler P., von Tucher S., Schieberle P. (2007) Influence of sulfur fertilization on the amounts of free amino acids in wheat. Correlation with baking properties as well as with 3-aminopropionamide and acrylamide generation during baking. *Journal of Agricultural and Food Chemistry, 55*, 4271-4277.

Graveland A., Bosveld P., Marseille J. P. (1978) Determination of thiol-groups and disulfide bonds in sheat-flour and dough. *Journal of the Science of Food and Agriculture, 29*, 53-61.

Grosch W. & Wieser H. (1999) Redox reactions in wheat dough as affected by ascorbic acid. *Journal of Cereal Science, 29*, 1-16.

Grove H., Hollung K., Moldestad A., Færgestad E. M., Uhlen A. K. (2009) Proteome changes in wheat subjected to different nitrogen and sulfur fertilizations. *Journal of Agricultural and Food Chemistry, 57*, 4250-4258.

Haneklaus S., Evans E., Schnug E. (1992) Baking quality and sulphur content of wheat. I. Influence of sulphur and protein concentration on loaf volume. *Sulphur in Agriculture, 16*, 31-36.

Haneklaus S., Murphy D. P. L., Nowak G., Schnug E. (1995) Effects of the timing of sulfur application on grain yield and yield components of wheat. *Zeitschrift für Pflanzenernährung und Bodenkunde, 158*, 83-85.

Hawkesford M. J., Davidian J. C., Grignon C. (1993) Sulfate proton cotransport in plasma-membrane vesicles isolated from roots of *Brassica napus* 1 - Increased transport in membranes isolated from sulfur-starved plants. *Planta, 190*, 297-304.

Hesse H. & Hoefgen R. (2003) Molecular aspects of methionine biosynthesis. *Trends in Plant Science, 8*, 259-262.

Huttner S. & Wieser H. (2001) Studies on the distribution and binding of endogenous glutathione in wheat dough and gluten. II. Binding sites of endogenous glutathione in glutenins. *European Food Research and Technology, 213*, 460-464.

Ismunadji M. & Dijkshoorn W. (1971) Nitrogen nutrition of rice plants measured by growth and nutrient content in pot experiments. Ionic balance and selective uptake. *Netherlands Journal of Agricultural Science, 19*, 223-236.

Jahn-Deesbach W. & Weipert D. (1964) Untersuchungen über den Einfluß der Stickstoffdüngung auf Ertrag und backereitechnologische Qualitätseigenschaften des Weizens. *Landwirtschaftliche Forschung, 18*, 132-145.

Kasarda D. D., Nimmo C. C. and Kohler G. O. (1978) Proteins and the Amino Acid Compositon of Wheat Fractions. In Ed. Y. Pomeranz, *Wheat Chemistry and Technology* (p 227). St.Paul, USA: American Association of Cereal Chemists, Inc.

Kettlewell P. S., Griffiths M. W., Hocking T. J., Wallington D. J. (1998) Dependence of wheat dough extensibility on flour sulphur and nitrogen concentrations and the influence of foliar-applied sulphur and nitrogen fertilisers. *Journal of Cereal Science, 28*, 15-23.

Kieffer R., Kim J. J., Walther C., Laskawy G., Grosch W. (1990) Influence of glutathione and cysteine on the improver effect of ascorbic acid stereoisomers. *Journal of Cereal Science, 11*, 143-152.

Larsson C. M., Larsson M., Purves J. V., Clarkson D. T. (1991) Translocation and cycling through roots of recently absorbed nitrogen and sulfur in wheat (*Triticum aestivum*) during vegetative and generative growth. *Physiologia Plantarum, 82*, 345-352.

Lawrence G. J. & Payne P. I. (1983) Detection by gel-electrophoresis of oligomers formed by the association of high molecular-weight glutenin protein subunits of wheat endosperm. *Journal of Experimental Botany, 34*, 254-267.

Luo C., Branlard G., Griffin W. B., McNeil D. L. (2000) The effect of nitrogen and sulphur fertilisation and their interaction with genotype on wheat glutenins and quality parameters. *Journal of Cereal Science, 31*, 185-194.

Macnicol P. K. (1977) Synthesis and interconversion of amino acids in developing cotyledons of pea (*Pisum-Sativum* L.). *Plant Physiology, 60*, 344-348.

MacRitchie F. (1980) Studies of gluten protein from wheat flours. *Cereal Foods World, 25*, 382-385.

Martre P., Porter J. R., Jamieson P. D., Triboi E. (2003) Modeling grain nitrogen accumulation and protein composition to understand the sink/source regulations of nitrogen remobilization for wheat. *Plant Physiology, 133*, 1959-1967.

Max Rubner-Institute (2008) Nationale Verzehrsstudie II. Ergebnisbericht, Teil 2; http://www.was-esse-ich.de/.

Miflin B. J. & Habash D. Z. (2002) The role of glutamine synthetase and glutamate dehydrogenase in nitrogen assimilation and possibilities for improvement in the nitrogen utilization of crops. *Journal of Experimental Botany, 53*, 979-987.

Moonen J. H. E., Scheepstra A., Graveland A. (1985) Biochemical-properties of some high molecular-weight subunits of wheat glutenin. *Journal of Cereal Science, 3*, 17-27.

Moss H. J., Wrigley C. W., MacRitchie F., Randall P. J. (1981) Sulfur and nitrogen fertilizer effects on wheat. 2. Influence on grain quality. *Australian Journal of Agricultural Research, 32*, 213-226.

Noctor G. & Foyer C. H. (1998) Ascorbate and glutathione: keeping active oxygen under control. *Annual Review of Plant Physiology and Plant Molecular Biology, 49*, 249-279.

Osborne T.B. (1924) The vegetable proteins. Longmans, Breen & Co, London.

Pelshenke P. F., Schulz A. and Stephan H. (1978) Rapid-Mix-Test, Standard-Backmethode für die Weizen- und Weizenmehlbeurteilung. In Ed. Arbeitsgemeinschaft Getreideforschung e.V., *Merkblatt der Arbeitsgemeinschaft Getreideforschung e.V.* (. Detmold: Granum-Verlag.

Popineau Y., Cornec M., Lefebvre J., Marchylo B. (1994) Influence of high M_r glutenin subunits on glutenin polymers and rheological properties of glutens and gluten subfractions of near-isogenic lines of wheat Sicco. *Journal of Cereal Science, 19*, 231-241.

Prieto J. A., Kelfkens M., Weegels P. L., Hamer R. J. (1992) Variations in the gliadin pattern of flour and isolated gluten on nitrogen application - implications for baking potential and rheological properties. *Zeitschrift für Lebensmittel-Untersuchung und Forschung, 194*, 337-343.

Randall P. J., Spencer K., Freney J. R. (1981) Sulfur and nitrogen fertilizer effects on wheat. 1. Concentrations of sulfur and nitrogen and the nitrogen to sulfur ratio in grain, in relation to the yield response. *Australian Journal of Agricultural Research, 32*, 203-212.

Ravanel S., Gakiere B., Job D., Douce R. (1998) The specific features of methionine biosynthesis and metabolism in plants. *Proceedings of the National Academy of Sciences of the United States of America, 95*, 7805-7812.

Reinbold J., Rychlik M., Asam S., Wieser H., Koehler P. (2008) Concentrations of total glutathione and cysteine in wheat flour as affected by sulfur deficiency and correlation to quality parameters. *Journal of Agricultural and Food Chemistry, 56*, 6844-6850.

Rennenberg H. (1982) Glutathione metabolism and possible biological roles in higher-plants. *Phytochemistry, 21*, 2771-2781.

Rennenberg H. (2001) Glutathione - An Ancient Metabolite with Modern Tasks. In D. Grill, M. Tausz and L. J. De Kok (Eds.), *Significance of Glutathione in Plant Adaptation to the Environment* (pp. 1-11). Dordrecht, The Netherlands: Kluwer Academic Publishers.

Rennenberg H., Schmitz K., Bergmann L. (1979) Long-distance transport of sulfur in *Nicotiana tabacum*. *Planta, 147*, 57-62.

Saito K. (2000) Regulation of sulfate transport and synthesis of sulfur-containing amino acids. *Current Opinion in Plant Biology, 3*, 188-195.

Schaafsma G. (2000) The protein digestibility-corrected amino acid score. *Journal of Nutrition, 130*, 1865S-1867S.

Schiff J. A. & Saidha T. (1987) Overview - Inorganic sulfur and sulfate activation. *Methods in Enzymology, 143*, 329-334.

Schröder P., Scheer C. E., Diekmann F., Stampfl A. (2007) How plants cope with foreign compounds - translocation of xenobiotic glutathione conjugates in roots of barley (*Hordeum vulgare*). *Environmental Science and Pollution Research, 14*, 114-122.

Seilmeier W., Belitz H. D., Wieser H. (1991) Separation and quantitative-determination of high-molecular-weight subunits of glutenin from different wheat-varieties and genetic-variants of the variety Sicco. *Zeitschrift für Lebensmittel-Untersuchung und Forschung, 192*, 124-129.

Seling S., Weigelt W., Wissemeier A. H. (2006) Bedeutung der Schwefeldüngung für Ertrag und Qualität von Weizen. *Getreidetechnologie, 60*, 148-152.

Shewry P. R. (1995) Plant storage proteins. *Biological Reviews of the Cambridge Philosophical Society, 70*, 375-426.

Shewry P. R. (2007) Improving the protein content and composition of cereal grain. *Journal of Cereal Science, 46*, 239-250.

Shewry P. R. (2009) Wheat. *Journal of Experimental Botany, 60*, 1537-1553.

Shewry P. R. & Halford N. G. (2002) Cereal seed storage proteins: structures, properties and role in grain utilization. *Journal of Experimental Botany, 53*, 947-958.

Shewry P. R., Halford N. G., Tatham A. S. (1992) High-molecular-weight subunits of wheat glutenin. *Journal of Cereal Science, 15*, 105-120.

Shewry P. R., Popineau Y., Lafiandra D., Belton P. (2000) Wheat glutenin subunits and dough elasticity: findings of the EUROWHEAT project. *Trends in Food Science & Technology, 11*, 433-441.

Shewry P. R. & Tatham A. S. (1997) Disulphide bonds in wheat gluten proteins. *Journal of Cereal Science, 25*, 207-227.

Shewry P. R., Tatham A. S., Halford N. G. (2001) Nutritional control of storage protein synthesis in developing grain of wheat and barley. *Plant Growth Regulation, 34*, 105-111.

Shewry P. R., Underwood C., Wan Y. F., Lovegrove A., Bhandari D., Toole G., Mills E. N. C., Denyer K., Mitchell R. A. C. (2009) Storage product synthesis and accumulation in developing grains of wheat. *Journal of Cereal Science, 50,* 106-112.

Simpson R. J., Lambers H., Dalling M. J. (1983) Nitrogen redistribution during grain-growth in wheat (*Triticum aestivum* L). 4. Development of a quantitative model of the translocation of nitrogen to the grain. *Plant Physiology, 71,* 7-14.

Statistisches Bundesamt (2010) *https://www-genesis.destatis.de/genesis/online;jsessionid=BAA09C17B0BC21CF3C72E94F81454E77.tomcat_GO_1_2?operation=abruftabelleAbrufen&selectionname=41241-0003&levelindex=1&levelid=1282214538094&index=2.* (18-9-2010)

Steinfurth D., Zörb C., Braukmann F., Mühling K.H. (submitted) Time-dependent distribution of sulphur, sulphate and glutathione in wheat tissues and grain as affected by late sulphur fertilization.

Tabe L., Hagan N., Higgins T. J. V. (2002) Plasticity of seed protein composition in response to nitrogen and sulfur availability. *Current Opinion in Plant Biology, 5,* 212-217.

Tea I., Genter T., Naulet N., Marie L. M., Kleiber D. (2007) Interaction between nitrogen and sulfur by foliar application and its effects on flour bread-making quality. *Journal of the Science of Food and Agriculture, 87,* 2853-2859.

Tea I., Genter T., Naulet N., Morvan E., Kleiber D. (2003) Isotopic study of post-anthesis foliar incorporation of sulphur and nitrogen in wheat. *Isotopes in Environmental and Health Studies, 39,* 289-300.

Tea I., Genter T., Violleau F., Kleiber D. (2005) Changes in the glutathione thiol-disulfide status in wheat grain by foliar sulphur fertilization: consequences for the rheological properties of dough. *Journal of Cereal Science, 41,* 305-315.

Triboi E., Abad A., Michelena A., Lloveras J., Ollier J. L., Daniel C. (2000) Environmental effects on the quality of two wheat genotypes: 1. quantitative and qualitative variation of storage proteins. *European Journal of Agronomy, 13,* 47-64.

Triboi E. & Triboi-Blondel A. M. (2002) Productivity and grain or seed composition: a new approach to an old problem - invited paper. *European Journal of Agronomy, 16,* 163-186.

van Beusichem M. L., Kirkby E. A., Baas R. (1988) Influence of nitrate and ammonium nutrition on the uptake, assimilation, and distribution of nutrients in *Ricinus communis. Plant Physiology, 86,* 914-921.

Vestreng V., Myhre G., Fagerli H., Reis S., Tarrason L. (2007) Twenty-five years of continuous sulphur dioxide emission reduction in Europe. *Atmospheric Chemistry and Physics, 7,* 3663-3681.

Wieser H., Gutser R., von Tucher S. (2004) Influence of sulphur fertilisation on quantities and proportions of gluten protein types in wheat flour. *Journal of Cereal Science, 40,* 239-244.

Wieser H. & Seilmeier W. (1998) The influence of nitrogen fertilisation on quantities and proportions of different protein types in wheat flour. *Journal of the Science of Food and Agriculture, 76,* 49-55.

Wrigley C. W., Ducros D. L., Fullington J. G., Kasarda D. D. (1984) Changes in polypeptide composition and grain quality due to sulfur deficiency in wheat. *Journal of Cereal Science, 2,* 15-24.

Wuest S. B. & Cassman K. G. (1992) Fertilizer-nitrogen use efficiency of irrigated wheat .1. Uptake efficiency of preplant versus late-season application. *Agronomy Journal, 84,* 682-688.

Yin L. P., Li P., Wen B., Taylor D., Berry J. O. (2007) Characterization and expression of a high-affinity nitrate system transporter gene (TaNRT2.1) from wheat roots, and its evolutionary relationship to other NTR2 genes. *Plant Science, 172,* 621-631.

Zhao F. J., Hawkesford M. J., McGrath S. P. (1999a) Sulphur assimilation and effects on yield and quality of wheat. *Journal of Cereal Science, 30,* 1-17.

Zhao F. J., Salmon S. E., Withers P. J. A., Evans E. J., Monaghan J. M., Shewry P. R., McGrath S. P. (1999b) Responses of breadmaking quality to sulphur in three wheat varieties. *Journal of the Science of Food and Agriculture, 79,* 1865-1874.

Zhao F. J., Withers P. J. A., Evans E. J., Monaghan J., Salmon S. E., Shewry P. R., McGrath S. P. (1997) Sulphur nutrition: an important factor for the quality of wheat and rapeseed (Reprinted from Plant nutrition for sustainable food production and environment, 1997). *Soil Science and Plant Nutrition, 43,* 1137-1142.

Zörb C., Grover C., Steinfurth D., Mühling K. H. (2010) Quantitative proteome analysis of wheat gluten as influenced by N and S nutrition. *Plant and Soil, 327,* 225-234.

Zörb C., Steinfurth D., Seling S., Langenkämper G., Koehler P., Wieser H., Lindhauer M. G., Mühling K. H. (2009) Quantitative protein composition and baking quality of winter wheat as affected by late sulfur fertilization. *Journal of Agricultural and Food Chemistry, 57,* 3877-3885.

Chapter 3

FUNCTIONAL GLUTEN ALTERNATIVES

Aleksandra Torbica, Miroslav Hadnađev,
Tamara Dapčević Hadnađev and Petar Dokić
Institute for Food Technology, University of Novi Sad, Novi Sad, Serbia

ABSTRACT

The demand for gluten-free products is ascending steadily, paralleling the increase in prevalence and incidence of celiac disease and other allergic reactions or intolerances to gluten consumption. The replacement of gluten presents a major technological challenge, as its visco-elastic properties largely determine the breadmaking performance of wheat flour. Various gluten-free formulations apply mixtures of rice or corn flour and different hydrocolloids or starches to mimic the unique properties of gluten. However, gluten-free products containing gums and starches as gluten replacements lack in essential nutrients. Therefore, replacing standard gluten-free formulations with those made from alternative grains like buckwheat may impart nutritional benefits. This chapter reviews the literature on gluten-free bakery products, including both hydrocolloid-based and pseudo-cereal-based formulations. A special emphasis is given to ongoing research in our laboratory related to gluten-free bread and cookies containing rice and buckwheat flour. The optimal gluten-free formulations were created by comparing the rheological properties of different rice and buckwheat flour mixtures to properties of wheat flour assessed by using Mixolab. Subsequently, rice, buckwheat and wheat flours were evaluated by electrophoretic and electron-microscopic analysis. Moreover, the influence of buckwheat flour type and content on rheological, textural, sensory properties of gluten-free dough, bread and cookies was investigated.

INTRODUCTION

As it is already known, wheat proteins are classified into gluten proteins (approximately about 80-85% of total wheat protein) and non gluten proteins (about 15-20% of total wheat proteins). Gluten proteins are consisted of prolamins (in wheat - gliadins) and glutelins (in wheat - glutenins) and non gluten proteins of albumins and globulins (Veraverbeke & Delcour, 2002).

Wheat storage proteins (glutens) have unique property of viscoelastic dough forming when wetted and kneaded. Such hydrated viscoelastic matrix of gluten proteins is also able to retain gas which is produced during the fermentation process resulted in getting aerated sponge-like structure of bread crumb (Stauffer, 2007; Marco & Rossel, 2008a). This ability can not be accompanied to any other known cereal flour, which makes wheat the most appropriate cereal for breadmaking (Marco & Rosell, 2008a). Moreover, the changes in glutenin–gliadin complex of grain have the great impact on technological quality of the wheat variety, thus affecting quality of the final product (Torbica et al., 2007).

According to Murray (1999), celiac disease represents life-long intolerance to the gliadin fraction of wheat and the prolamins or rye (secalins), barley (hordeins) and possibly oats (avidin). Due to its adverse affect on health of celiac sufferers, gluten must be eliminated from the diet which is very demanding task because of its structure-forming ability which affects elastic properties of dough and contributes to the overall appearance and crumb structure of many baked products. Therefore, the removal of gluten in gluten-free formulation very often results in low quality, crumbling texture, pour colour, and mouthfeel and low flavour products. Also, utilization of different gluten-free flours as raw materials is limited due to the need of the changing of traditional production process, making it difficult challenge for cereal technologists and to bakers (Gallagher et al., 2004; Marconi & Careca 2001).

Due to the fact that wheat gluten possesses wide varieties of roles in breadmaking, the production of gluten-free products represents serious task. Therefore, in order to achieve a good quality product which is also simultaneously gluten-free, it is necessary to use wide range of ingredients. One of the disadvantages, flaws, of many gluten-free flours i.e. gluten-free products is that they are wheat based which can therefore endanger the health of patients suffering from celiac disease, because of the facts that some residues of gluten might be still present. Also, the excessive usage of gluten-free products made of wheat or other gluten containing cereal sources might have negative health effect on celiac patients. According to Codex Alimentarius definition, gluten-free foods are foodstuffs: a) consisting of or made only from ingredients which do not contain any prolamins from wheat or all Triticum species such as spelt (*Triticum spelta L.*), kamut (*Triticum polonicum L.*) or durum wheat, rye, barley, [oats] or their crossbred varieties with a gluten level not exceeding [20 ppm]; or b) consisting of ingredients from wheat, rye, barley, oats, spelt or their crossbred varieties, which have been rendered "gluten-free"; with a gluten level not exceeding [200 ppm]; or c) any mixture of the two ingredients as in a) and b) with a gluten level not exceeding [200 ppm].

In past years serious attention was dedicated to substitution of gluten by ingredients which are able to imitate gluten properties (Sciarini et al., 2008). However, large numbers of market available gluten-free breads are poorer quality in comparison to their gluten-containing counterparts (Gallaghe et al., 2003). The large groups of gluten-free cereal foods are based on refined gluten-free flour or starch and they are not nutritionally enriched or fortified (Thompson, 1999). In order to develop good quality gluten-free product, it is recommended to use a range of flours and polymeric substances that are capable of mimicking the viscoelastic properties of gluten. Therefore, the combination of different gluten-free flours rather than just one gluten-free flour type have to be considered for obtaining gluten-free product with good sensory and textural properties (Arendt et al., 2008).

Also, during past years there is increasing number of papers referring to gluten-free breads produced by incorporation of different starches, dairy proteins and hydrocolloids into gluten-free flours in order to improve overall rheological an sensory properties as well as the shelf life of such obtained products (Moore et al., 2004; Ahlborn et al., 2005; Gallagher et al., 2003; Gallagher et al., 2004; Haque & Morris, 1994; McCarthy et al., 2005; Moore et al., 2006; Schober et al., 2005; Sivaramakrishnan et al., 2004; Toufeili et al., 1994; Ylimaki et al., 1991).

Rice flour is widely used as gluten-free raw material due to its soft taste, hypoallergenic properties, low levels of sodium and easy digestible carbohydrates, it is colourless and it represents one of the most abundant cereal crops in the world. All these properties influenced that rice flour is the most suitable cereal in gluten-free industry (Gujral et al., 2003a; Gujral et al., 2003b; Gujral & Rosell, 2004a, 2004b; Lopez et al., 2004).

Since rice flour proteins are mainly consisted of glutelins (65-85%) and only small amount of prolamins (2.5 - 3.5%) is present, rice flour during hydration and kneading is not able to form network with properties similar to wheat gluten properties (Marco and Rosell, 2008a; Gujral & Rosell, 2004b). Also, protein network built by rice flour proteins is not able to retain CO_2 produced during the fermentation process. Therefore, in order to improve viscoelastic properties, different structuring agents such as hydroxypropylmethylcellulose (HPMC), locust bean gum, guar gum, carrageenan, xanthan gum, agar, (Kang et al., 1997) pectin, carboxymethylcellulose (CMC), agarose and oat b-glucan (Lazaridou et al., 2007) emulsifiers, enzymes and dairy products (Demirkesen et al. 2010) etc. were used in combination with rice flour. Rice flour is mainly used as milled rice. Prior the milling process husk and the bran is removed and also the fiber, vitamins and other nutritional components making it poor in fibers and nutrients (Rosell & Marco, 2008). Therefore, the combination of rice flour with different other flours might lead to a gluten-free product with improved nutritional properties. For example, it is well known that soybean flour is a good counterpart for cereal flours due to presence of lysine and methionine amino acids in soybean flour (Iqbal et al., 2006). Also, the addition of different source proteins such as egg albumen and dairy proteins could improve nutritional properties of gluten-free products (Gallagher et al., 2003, Ribotta et al., 2004; Moore et al, 2006; Marco & Rosell, 2008b). In past years much more attention is dedicated to alternatives for gluten containing grains. This large group comprises the pseudocereals which include amaranth, quinoa and buckwheat that are all nutrient-dense and their incorporation in gluten-free formulations will improve the nutritional quality of these products (Kupper, 2005; Alvarez-Jubetea et al., 2010; Schoenlechner et al., 2008).

Generally, two types of buckwheat are cultivated around the world: common buckwheat (*Fagopyrum escelentum* Moench) grown in Europe, USA, Canada, Brazil, South Africa and Australia and tartary buckwheat (*Fagopyrum tataricum*) cultivated in mountainous regions of southwest China (Bonafacciaa et al., 2003; Lin et al., 1992).

The proteins found in buckwheat are mainly consisted of globulins and albumins and very low content of storage prolamin proteins. The prolamins of cereals are main storage proteins and they are labelled as the toxic proteins in celiac disease (Drzewiecki et al., 2003; Gorinstein et al., 2002). Furthermore, amino acid composition of buckwheat proteins is well balanced containing higher percentage of essential amino acids such is lysine making pseudocereals protein composition superior to cereals composition (Aubrecht & Biacs, 2001; Drzewiecki et al., 2003; Gorinstein et al., 2002). Also, positive effect of buckwheat intake on human health can be observed in lowering serum cholesterol and in higher ratio of high density lipoprotein cholesterol to total cholesterol (He et al., 1995). Buckwheat is a rich source of calcium, magnesium, iron and other minerals (Alvarez-Jubete et al., 2009) making it attractive for gluten-free formulations that are known to be deficient in these minerals (Hopman et al., 2006; Kupper, 2005; Thompson, 2000; Thompson et al.,2005). According to Bonafaccia et al., buckwheat possesses good content of thiamine, riboflavin and pyridoxine. Buckwheat contains large amounts of flavoniods or/and polyphenols (Watanabe et al., 1997). Rutin and quercetin represent the major group of polyphenol with antioxidative activity found in buckwheat (Oomah, B. D., & Mazza, G.,1996; Holasova et al., 2002). It was also found that buckwheat seeds contain higher amount of dietary fiber compared to other pseudocereals i.e. amaranth and quinoa (Alvarez-Jubete et al., 2009). The consumption of buckwheat could have positive impact on celiac patients diet due to deficiency of dietary fiber intake in common celiac disease diet based on refined flours and starches (Alvarez-Jubete et al.,2010).

However, the creation of gluten-free product which will be, at the same time, nutritionally improved and rheologically similar to wheat flour product is a very challenging task for many cereal science researchers and bakers. The reason is that wheat dough is probably the most dynamic and complicated rheological system, and its rheological assessment is very important, since it can provide numerous information about dough formulation, structure and processing. Rheological measurements are being used as indicators of the gluten polymer molecular structure and predictors of its functional behaviour in breadmaking (Dobraszczyk & Morgenstern, 2003). Additionally the rheological properties of dough can be indicative of the quality of the finished product. Determination of bread rheological properties has its roots in long tradition of subjective manual assessments of dough rheology prior to baking, e.g. kneading and stretching the dough by hand to assess its quality.

Rheology, as a science, does not have the application only in cereal science, but also contribute to a number of applications such as colloids, suspensions and emulsions (Djaković & Dokić, 1978; Dokić-Baucal et al., 2004; Dokić et al., 2008; Krstonošić et al., 2009), synthetic and biopolymers (Djaković & Dokić, 1972; Dokić & Djaković, 1975; Dokić et al., 1998; Dokić et al., 2010). From rheological point of view dough is a viscoelastic material with nonlinear behaviour. It exhibits shear thinning, meaning that at low strains viscosity is high and dough structure intact, while at high strains viscosity is reduced due to dough structure disorientation and destruction. Moreover, wheat dough exhibit thixotropy which was described by Muller (1975).

In this research rheological assessment of dough was performed by using empirical and fundamental methods. Mixolab measurements were used as a tool for evaluation of the possibility of different cereals, pseudocereals and other raw materials to mimic unique rheological behaviour of wheat gluten. In order to study the structural aspects responsible for similar behaviour of rice/buckwheat flour blends and wheat flour, scanning electron microscopy and electrophoresis were used for the integration of the information coming from empirical rheological measurements. Subsequently, the optimal rice/buckwheat flour blends were used to prepare gluten-free bread and cookies which were investigated using fundamental rheological, textural and sensory evaluation.

MATERIALS AND METHODS

Materials

Rice flour, RF (moisture content 9.09%, protein 7.31%, ash 0.26%, cellulose 0.16%, lipid content 0.30% and starch 81.5%), unhusked buckwheat flour, UBF (moisture content 9.76%, protein 12.38%, ash 2.19%, cellulose 3.02%, lipid content 2.77% and starch 67.38%) and husked buckwheat flour, HBF (moisture content 10.11%, protein 7.50%, ash 1.09%, cellulose 0.43%, lipid content 1.75% and starch 68.24%), as well as amaranth flour, soybean flour, wheat flour and corn flour were purchased from Hemija Komerc, Novi Sad, Serbia. Sodium carboxymethylcellulose (CMC) was purchased from Centrohem, Beograd, Serbia. Sodium bicarbonate ($NaHCO_3$) was obtained from Zorka Pharma, Šabac, Serbia and DATEM - diacetyl tartaric acid esters of mono- and diglycerides of fatty acids (PANTEX DW 90) was obtained from Incopa, Germany. Vegetable fat-shortening was obtained from Puratos, Belgium while the other ingredients (salt, sugar, fresh yeast, honey) were purchased from local market.

Methods

Determination Of Optimal Gluten-Free Blend Using Mixolab

Mixing and pasting behaviour of the wheat flour was compared to behaviour of different cereal and pseudo cereal flours using the Mixolab (Chopin, Tripette et Renaud, Paris, France), which measures the torque (expressed in Nm) produced by passage of dough between the two kneading arms, thus allowing the study of its rheological behaviour. After determining the optimal gluten-free blend (having similar rheological behaviour to wheat flour) the Mixolab curves of those blends were also recorded. For the assays, proper amount (to obtain 90g of dough) of flour or flour blend was placed into the Mixolab analyser bowl and mixed. After tempering the solids, the water required for optimum consistency (1.1 Nm) was added. The settings used in the test were 8 min at 30 °C with a temperature increase of 4 °C/min until the mixture reached 90 °C; at this point, there was an 7 min holding period at 90 °C, followed by a temperature decrease of 4 °C/min until the mixture reached 50 °C, and then 5 min of holding at 50 °C. The mixing speed during the entire assay was 80 rpm. The measurement was repeated twice for each flour and blend.

Determination Of Properties Of Optimal Gluten-Free Blends

Since rice and buckwheat flour (husked and unhusked), as well as their blends were chosen as ones having similar rheological behaviour to wheat flour, they were used in all the other measurements. In rice/husked buckwheat flour and rice/unhusked buckwheat flour blends the ratio of rice flour to buckwheat flour was 90:10, 80:20 and 70:30. Further increase in buckwheat flour content was not performed due to possible negative effect of buckwheat flour on sensory properties of the final product.

Electrophoresis

Samples (30 mg) of wheat, husked buckwheat and rice flours as well as mixtures of rice and buckwheat flour (90:10; 80:20; 70:30) were placed in Eppendorf tubes and 1000 µL of extraction buffer (0.125 M TRIS-HCl pH 6.8, 4% SDS, 20% glycerol, 10% β-mercaptoethanol). The mixture was agitated on a rotary shaker for 30 min and then centrifuged at 14 000 g for 25 min. The supernatant was removed and the extract was used to prepare samples for the analysis as described below.

The chip-based separations were performed on the Agilent 2100 bioanalyzer (Agilent Technologies, Santa Clara, CA, USA) in combination with the Protein 230 Plus LabChip kit and the dedicated Protein 230 software assay on 2100 expert software. All chips were prepared according to the protocol provided with the Protein 230 LabChip kit. The channels of the chip have to be filled with a mixture of a sieving matrix (has non–cross-linked linear format) and a fluorescent dye for detection. The channels are filled by pipetting 12 µl of the gel dye mixture into one of the wells and applying pressure with a syringe for 60 s. The Protein 230 dye is a blue fluorescent dye that interacts with the protein SDS micelles. The gel–dye mixture is also added to the other system wells, where it serves as a buffer reservoir during the separation. The sample preparation is comparable to SDS-PAGE. Sample buffer (2 µl), including lithium dodecylsulfate; a reducing agent (if applicable); and two internal standards (lower and upper marker), are added to a 4-µl sample, and 84 µl deionised water.

The samples are heat denatured at 95–100 °C for 3–5 min before loading them onto the chip. The chip is then placed into the bioanalyzer. Once the chip is placed into the instrument, the electrodes touch the liquids in the well, forming an electric circuit. These electric circuits make it possible to move samples from the sample well into the channels and to perform injections into the separation channel. Each sample is sequentially separated in the separation channel and detected by laser-induced fluorescence detection (670–700 nm) within 45 sec. The complete analysis of six protein samples, including sizing and quantitation, takes 25 min (including the start-up phase of the instrument). After completion of the chip run, the software offers the alternatives of displaying results as quantitative profiles (as for conventional liquid chromatography) and also as simulated gel-electrophoresis patterns. Fractionation is size based, and the profiles show the smallest proteins emerging first in the profiles.

Scanning Electron Microscopy (SEM)

The surface morphology of the different flour samples was imaged using a scanning electron microscopy (SEM). Samples were coated with Au using a Spater coater device Baltec SCD 005. The micrographs are obtained using a SEM-Jeol JSM 6460LV instrument. The micrographs were taken by using magnification of 500 x.

Bread Formulations

Gluten-free blends were prepared by mixing rice flour and husked buckwheat flour or rice flour and unhusked buckwheat flour in ratios 90:10, 80:20 and 70:30 based on numerous previously conducted measurements.

The list of bread dough ingredients is presented in Table 1. The content of ingredients is expressed as g in 100 g gluten-free blend.

Table 1. Gluten-free bread formulations containing blends of rice flour with husked or unhusked buckwheat flour in ratios 90:10, 80:20 and 70:30

Ingredient	%
Gluten-free blend	100
Deionised water	180 or 190[A]
Fresh yeast	8
Sugar	4
Salt (NaCl)	3
Vegetable fat	4

[A] The amount of added water in formulations with HBF was 180%, while in UBF was 190% due to higher water absorption values observed using Mixolab measurements

For the breadmaking purpose all dry ingredients, vegetable fat and appropriate amount of water (previously tempered at 30 °C) were mixed with a 4-speed mixer (Gorenje MRP 275 EA) for 2 min. Consequently, fresh yeast was added and mixed for additional 2 min. 90 g of dough was poured into polytetrafluoroethylene moulds (diameter 6 cm) and proofed for 15 min at 30 °C. The baking tests were carried out at 220 °C for 35 min using laboratory baking oven (MIWE condo, Michael Wenz).

Cookie Formulations

The gluten-free blends containing rice flour and husked buckwheat flour in ratios 90:10, 80:20 and 70:30 were used to prepare gluten-free cookies and the ratio of blend to other cookie ingredients is listed in Table 2. The content of ingredients is expressed as g in 100 g gluten-free blend.

Table 2. Gluten-free cookie formulations containing blends of rice flour and husked buckwheat flour in ratios 90:10, 80:20 and 70:30

Ingredient	%
Gluten-free blend	100
Deionized water	35
Vegetable fat	33.33
Granulated sugar	25
Honey	15
Sodium bicarbonate (NaHCO$_3$)	3
DATEM	3
Carboxymethilcellulose (CMC)	1.5
Salt (NaCl)	0.7

Gluten-free cookies were produced according to commercial formulation and bakery experience. The ingredients were kept at 25 °C for 24 h. All powder materials were mixed together in Brabender Farinograph mixer at 30 °C. Flour or flour blend was added first and then the other powder materials (salt, NaHCO$_3$, granulated sugar, CMC) and mixed for a 3 minutes. Subsequently, the whole mass of shortening was added in the mixer and mixed for 2 minutes keeping the lid on. Finally, water tempered at 30 °C with dissolved appropriate amount of honey was poured in the mixer bowl. The mixing was kept in the period of 25 minutes at 30°C with periodical scraping of the dough.

The mixed dough was left to rest for 24 h at 20 °C in order to allow hydratation of the added carboxymethilcellulose. Subsequently, the dough was sheeted between two cylinders of a pilot scale sheeter (Sfogliatrice Mignon, Italy). The gap settings between the cylinders were: 14 mm – one passage, 10 mm – one passage, 7 mm – two passages, 5 mm – two passages and 4 mm – two passages with 15 s resting period between each passage. The dough was then cut using a mould (60 x 55 mm) and baked immediately at 170 °C in a laboratory oven (MIWE gusto® CS, Germany) for 12 minutes. The cookies were then cooled at room temperature for 2 h and afterwards they were packed in sealed bags.

Dynamic Oscillatory Measurements

Dynamic oscillatory measurements were performed using Haake Mars (Thermo Scientific, Germany). In the case of bread batter the test was conducted on dough consisted of all the ingredients given in the Table 1, but without the addition of the fresh yeast. However, the dynamic oscillatory tests for cookie dough were conducted immediately after the sheeting process. The measuring system for gluten-free bread consisted of parallel plate geometry (60 mm diameter, 1 mm gap) while for gluten-free cookies serrated plate geometry (35 mm diameter, 1 mm gap) was used in order to avoid slippage of sample during the measurements (increased amount of fat in cookies recipes in comparison to bread formulation). After loading, the dough sample was left to rest for 10 min, so that residual stresses could relax. Excess dough was trimmed just before the measurement and the exposed edges of the samples were covered with low viscosity paraffin oil to prevent moisture loss and drying during measurement.

Stress sweeps were performed at 25 °C and a frequency of 1 Hz for all samples to determine the linear viscoelasticity zone. Frequency sweeps tests, at 0.1 Pa and 25 °C were performed from 0.1 to 10 Hz. The oscillatory rheological parameter used to compare the viscoelastic properties of all gluten-free products was: elastic (storage) modulus (G').

All rheological measurements were performed in three replicates.

Textural Measurements and Sensory Evaluation of the Final Product

Textural properties of the final gluten-free products 9bread and cookies) were investigated using Texture analyzer TA.XPplus (Exponent Stable Micro System).

Bread firmness test was performed using modified AACC (74-09) method due to dimensions of final product which were smaller (bottom diameter 6 cm and height cca. 6 cm) than the standard bread loaf. Modification involved investigation of the textural properties of the whole product instead of the breadcrumb solely using P/0.5 inch diameter cylinder probe.

Textural measurements were performed 2 hours after baking and storage at room temperature and the obtained results were expressed as the hardness of the final product.

On contrary, cookie break strength was estimated using 3-Point Bending Rig (HDP/3PB) and 5 kg load cell in compression mode. Pre-test speed, test speed and post-test speed were 1.0 m/s, 3.0 mm/s and 10.0 mm/s, respectively. The gap distance of adjustable support of the base plate was 55 mm and upper blade was in equidistant position from them. The prior adjustments were kept constant during all the measurements. Maximum force recordings during the measurements can be referred to as cookie break strength. Textural analyses were conducted after 24 h. All textural measurements were conducted at 20 °C in nine replicates per batch.

In order to evaluate differences between cookie formulations sensory evaluation was performed. It was conducted by a group of 10, both male and female, trained panellists who were familiar with sensory analysis techniques (Resurrecion, 2008). The following sensory attributes were evaluated: appearance, taste and flavour. The sensory analysis was carried out 2 h after bread baking and 24 h after cookie baking.

Statistical Analysis

A one-way analysis of variance and Tukey's test were used to establish the significance of differences among the mean values at the 0.05 significance level. The statistical analyses were performed using Statistica 8.0 (Statsoft, Tulsa, USA).

RESULTS AND DISCUSSION

Mixolab Assessment and Characterization of Optimal Gluten-Free Blends

Mixolab, as rheological instrument, allows analysis of the quality of the protein network and the starch behaviour during heating and cooling, thus simulating the breadmaking process during mixing, baking and cooling. Therefore, Mixolab can be used as suitable tool in creating new product formulations. In this work Mixolab has been used to investigate in which manner non gluten containing cereals (rice, corn), pseudocereals (amaranth, buckwheat) and other raw material (soybean) can be used to mimic wheat flour rheological properties (Figure 1). Among all tested materials the mixtures of RF and HBF as well as RF and UBF were selected since their Mixolab curves were more similar to the wheat flour curve than the curves obtained for other flours such as soybean, amaranth and corn.

Moreover, observing the second part of the curve it can be noticed that wheat flour curve is positioned between the curves of RF and BF, meaning that mixtures of RF and BF would perform the most optimal properties during the heating period.

The obtained Mixolab parameters of RF/UBF mixtures and RF/HBF mixtures containing 90, 80 and 70% of rice flour, as well as the data of wheat flour, rice flour and husked and unhusked buckwheat flours are summarized in Table 3.

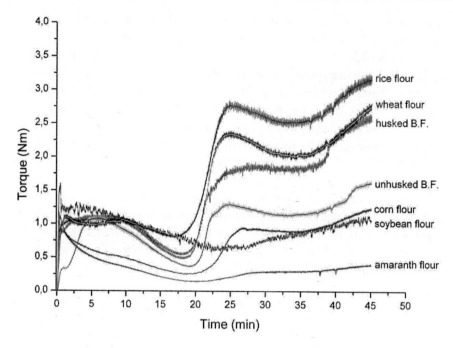

Figure 1. Mixolab profiles of rice flour, wheat flour, husked and unhusked buckwheat flour, amaranth flour and corn and soybean flour.

Table 3. Mixolab parameters of rice flour, husked and unhusked buckwheat flour, wheat flour as well as the gluten-free mixtures obtained from rice flour (RF) and husked buckwheat (HF) or unhusked buckwheat (UBF) flour [A]

	Development Time (min)	Stability Time (min)	C2 torque (Nm)	C3 torque (Nm)	C5-C4 torque (Nm)	W abs (%)
Rice flour	8.76[g]	12.22[f]	0.81[f]	2.77[j]	0.64[bc]	61.7[d]
HBF	6.63[f]	10.96[de]	0.49[c]	1.86[b]	0.77[d]	58.1[a]
UBF	5.93[e]	5.66[a]	0.37[a]	1.30[a]	0.49[a]	67.7[f]
Wheat flour	1.35[a]	11.78[ef]	0.53[c]	2.31[eg]	0.69[c]	65.4[e]
90%RF+10%HBF	1.38[a]	11.61[ef]	0.72[e]	2.56[i]	0.86[e]	59.1[b]
80%RF+20%HBF	1.63[a]	10.25[cd]	0.59[d]	2.49[h]	0.79[d]	59.3[b]
70%RF+30%HBF	2.48[b]	10.28[cd]	0.52[c]	2.36[g]	0.81[de]	59.0[b]
90%RF+10%UBF	2.60[bc]	9.66[c]	0.61[d]	2.29[e]	0.82[de]	60.5[c]
80%RF+20%UBF	2.98[c]	8.00[b]	0.53[c]	2.09[d]	0.79[d]	60.6[c]
70%RF+30%UBF	3.75[d]	7.72[b]	0.43[b]	1.93[c]	0.63[b]	61.7[d]

[A]Values represent the means. Values followed by different lower-case letters in the same column are significantly different from each other ($p \leq 0.05$).

The first part of the Mixolab curve refers to the protein characteristics of the systems and it is characterized by the following parameters: water absorption (W abs); dough development time; dough stability and C2 value which is related to the protein weakening due to mechanical and thermal constraints. As it can be seen in Table 3, development times of RF, UBF, HBF and WF were significantly different. Observing the RF/HBF and RF/UBF systems it can be seen that systems containing 10% and 20% of HBF were not significantly different than WF dough concerning dough development time. However, the increase in the amount of both types of BF and especially UBF resulted in longer dough development time due to the increase in fiber (cellulose) content which requires longer period of time to absorb water. According to the values presented in Table 3 it can be concluded that the mixture containing UBF had longer development time due to higher fiber content. Increasing the content of HBF from 10% to 20%, stability of the system decreased having no additional impact on dough stability with further HBF addition (30%). More expressed impact on stability can be observed for the system consisted of UBF and RF (Table 3). Due to higher content of fiber materials present in UBF compared to HBF, those systems had higher water absorption values and lower stability which decreased by increasing the amount of UBF in tested systems. Significant difference between C2 values can be seen by comparing the RF, HBF and UBF. The presence of rice flour in the tested systems was dominant factor which affected the increase in C2 values. Therefore, mixtures containing 10% and 20% of HBF and 10% of UBF possessed significantly higher value of C2 than the WF system. For both systems (RF/HBF and RF/UBF), increasing the buckwheat flour content resulted in more intensive weakening of protein network. It can be assumed that buckwheat flour possesses much worse protein quality from the technological point (Mariotti et al., 2008) of view than the rice flour and therefore the addition of buckwheat flour resulted in decrease in protein dependent baking characteristics responsible for the final product properties. The addition of UBF, which has higher cellulose content than HBF, resulted in larger decrease in C2 value in comparison to the systems with HBF.

Second part of the Mixolab curves reveals the starch properties of tested systems. More attention was paid on the first peak at C3 point which is the measure of starch gelatinization and the difference between the C5 and C4 value which represents starch retrogradation degree (Bonet et al., 2006; Ozturk et al., 2008). It can be seen (Table 3) that RF starch behaviour was characterized by the highest gelling ability as it was manifested by the value of C3 point. Increasing the amount of both HBF and UBF in tested dough samples, the values of torque at C3 point and the value of C5-C4 decreased. All the values at C3 point for the systems with HBF were higher in comparison to the systems containing UBF due to lower protein and lipid content in HBF. Namely, it has been reported that lower lipid and protein content is associated with a higher peak viscosity, indicative of higher starch swelling (Debet & Gidley, 2006; Nelles et al., 2000). It was also observed that difference between C5 and C4, which is correlated to starch retrogradation degree, decreased with the increase of buckwheat flour, especially UBF, which indicates that the addition of BF to the bread or other bakery product could lead to the product with improved anti-staling properties.

In order to identify structural properties responsible for similar behaviour of rice and buckwheat flour dough to wheat flour dough electrophoretic and SEM measurements were conducted.

Electrophoresis

Electrophoregrams of endosperm proteins of wheat flour, rice flour and buckwheat flours are shown in Figure 2.

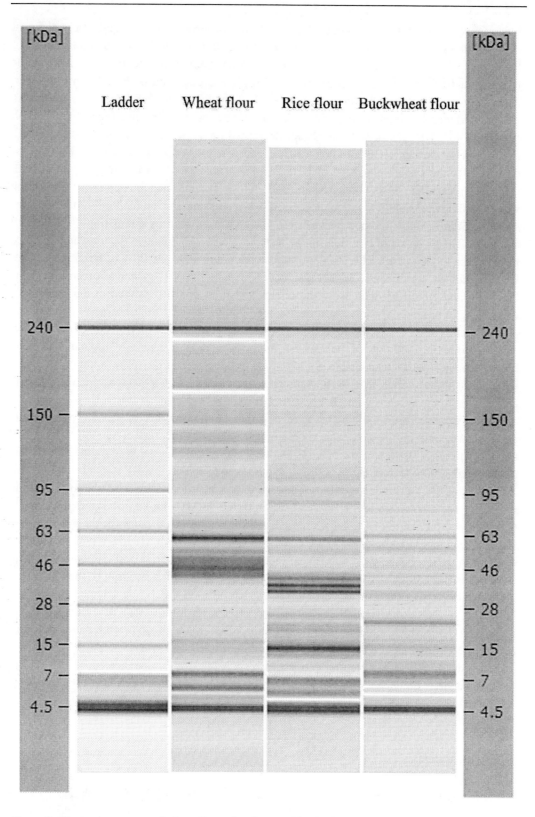

Figure 2. Electrophoregrams of wheat flour, rice flour and buckwheat flours.

Qualitative and quantitative differences between WF, RF and BF electrophoregrams can be observed. Protein fraction of WF were found to be in the range of molecular weight of 10 – 200 kDa, proteins of RF in the range of molecular weight between 7 – 100 kDa and BF proteins between 10 – 90 kDa.

Observing the overlaying LoaC profiles of RF and BF (Figure 3) evident differences between content of extracted proteins were registered by Expert Software 2100 (Agilent Technologies, Palo Alto, CA). Total amount of extracted proteins of rice flour was higher in comparison to buckwheat flour. Due to this differences, by increasing the amount of BF in flour mixtures resulted in decreased of total extracted protein content (Figure 4). This was in contradiction with the expectations considering that RF and BF samples had almost the same protein content estimated by standard AOAC procedure.

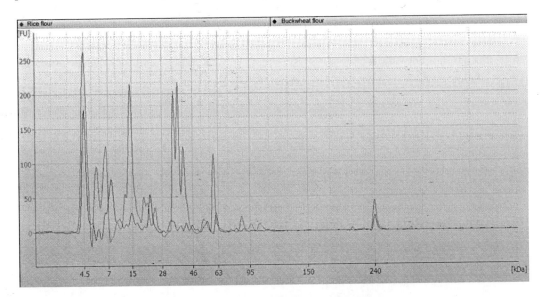

Figure 3. Electrophoretic LoaC profiles of rice flour and buckwheat flour.

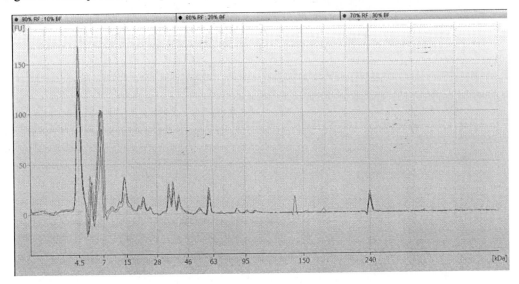

Figure 4. Electrophoretic LoaC profiles of rice/buckwheat flour blends.

Due to different structure and solubility of total as well as reserve proteins of both buckwheat seeds and rice grains, resulted in different protein extraction from the centre of the endosperm. The above mentioned difference between the amounts of extractible proteins originated from the fact that the nature of the reserve proteins of buckwheat and rice is very different. Reserve proteins of rice grains are prolamins and of buckwheat seeds are globulins. According to Ju et al. (2001) 83% or 87% of total rice proteins are glutelins and prolamins. The content of glutenin and prolamin in buckwheat seeds is found to be in the range of 43 - 45% of total protein content (Bejosano & Corke, 1999).

It can be observed from the electrophoretic LoaC results (Figure 3), that extracted proteins of rice flour and buckwheat flour possess similar physical-chemical characteristics due to the identical extraction conditions. Having on mind that the extraction conditions are adapted to optimal conditions of wheat gluten complex extraction, obtained results imply that extracted rice and buckwheat flour proteins have similar properties as the one of gliadin and glutenin of wheat flour. This means that in dough formation conditions these proteins are able to form a protein network which can be responsible for structural and mechanical properties of the final product containing these two types of flours.

Scanning Electron Microscopy

The SEM images of flour samples are shown in Figure 5.

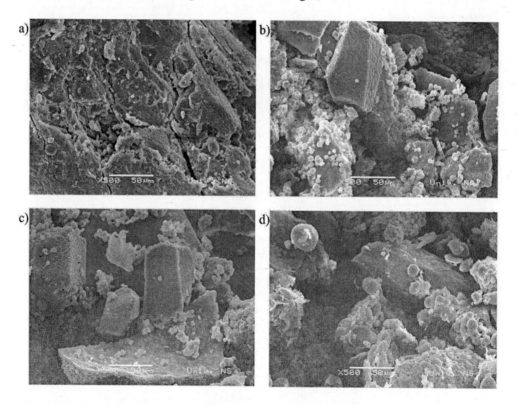

Figure 5. SEM images of rice (a), husked buckwheat (b), unhusked buckwheat (c) and wheat (d) flour.

The resulting flours vary in particle size and differ in physical properties. The all flour samples particles are not spherical and their constituents have different size and linkage between each other. Rice flour has the smallest starch granules and the most compact structure in comparison to other three flours. Husked and unhusked buckwheat flours have a specific tightly lamellar block structure of particles. They have not visible protein network like wheat flour particles in Figure 5. This is probably consequence of different protein nature in these flours. Also, wheat flour particles in generally have the largest starch granules and wide range of starch granules size which are not packed so close as like as buckwheat; among them large and small granules could be distinguished. On contrary, at the cellular level, gliadins and glutenins look as an amorphous protein matrix deposited between starch granules in the endosperm cells (Bechtel & Wilson, 2003).

The Properties of Gluten-Free Products

According to previously conducted Mixolab, electrophoretic and SEM measurements two different gluten-free products (bread and cookies) were prepared. Consequently, they were characterised by rheological, textural, and sensory analyses.

Rheological Properties of Gluten-Free Bread

The viscoelastic behaviour of gluten-free dough samples was investigated by oscillation frequency sweep experiments conducted in the linear viscoelastic range. Elastic (G') modulus which represents stored deformation energy and viscous (G") modulus representing lost energy were measured. According to the obtained results, elastic modulus was significantly higher than viscous modulus through all the frequency range, showing only slight increase with increasing frequency, which is typical gel-like behaviour (Witczak et al., 2010). G' values of all gluten-free bread formulation, obtained from frequency sweep tests, are listed in Table 4.

Table 4. Values of storage modulus of gluten-free bread dough containing husked buckwheat (HF) or unhusked buckwheat (UBF) flour [A]

Type of buckwheat flour	Content of buckwheat flour (%)	G' (Pa)
HBF	10	4182±457.80[c]
	20	7165±351.61[d]
	30	1396±129.40[a]
UBF	10	8234±883.30[d]
	20	2876±168.49[b]
	30	2894±163.54[b]

[A]Values represent the means±standard deviation. Values followed by different lower-case letters are significantly different from each other ($p \leq 0.05$).

With increasing UBF concentration from 10% to 20% there was a decrease in G' value, indicating a decrease in elasticity of dough, while the further increase in UBF concentration to 30% did not influence the elasticity of sample.

On contrary, systems containing HBF have shown initial increase and subsequent decrease in G' and value with increasing the amount of buckwheat flour. It seems possible that changes in gel strength were the consequence of changes in macromolecular organisation due to segregative interactions between components of the system. Namely, it was already proven that interactions between protein and polysaccharide lead to changes in profile of mechanical spectrum (G' and G" values) (Fitzsimons et al., 2008; Harrington et al., 2009).

Textural and Sensory Analysis of Gluten-Free Bread

Textural properties of gluten-free bread were evaluated by measuring the work of compression which is common parameter in describing hardness properties of tested systems.

The obtained values of the works of compression for the products containing different proportions of RF and both buckwheat flours (Table 5) were not significantly higher for the samples prepared with UBF than for those containing HBF ($p > 0.05$).

Also, the resulted works of compression increased but not significantly ($p > 0.05$) with increasing the BF content in the final product. Therefore, the increase in BF addition did not significantly affect the textural properties of the final product. This observation indicated that gluten-free bread containing higher amount of nutritionally valuable buckwheat flour (Lin et al., 2009; Prestamo et al., 2003), could be produced without affecting the textural properties of the product.

Table 5. Textural properties of final gluten-free breads containing husked buckwheat (HF) or unhusked buckwheat (UBF) flour [A]

	Type of buckwheat flour	10%	20%	30%
Area (g.sec)	HBF	3261±190[a]	4266±297[ab]	4439±247[abc]
	UBF	4121±348[ab]	4706±403[bc]	5597±823[c]

[A] Values represent the means± standard deviation. Values in table followed by different lower-case letters are significantly different from each other ($p \leq 0.05$).

Upper bread crust and breadcrumb appearance are illustrated in Figure 6. As it can be seen the increase in the amount of BF resulted in cracked surfaces of the upper crust of gluten-free products. However, the increase in the amount of BF led to improved breadcrumb structure which was characterised with more uniform pore distribution.

Observing the taste properties it was estimated that increase in UBF content led to product with more bitter taste due to nonvolatile compounds responsible for a bitter taste which are predominantly found in the husk (Janeš et al., 2010). Unlike the UBF containing products, bread prepared with HBF was more tasteful, even when concentration of HBF was 30%. Moreover, products containing higher amount of HBF expressed more pleasant flavour and taste. Generally, all the gluten-free samples were found to be sensory acceptable.

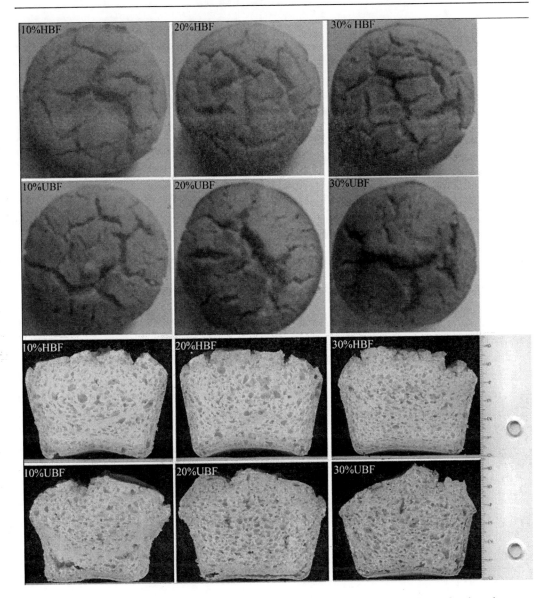

Figure 6. Upper surface crust appearance and breadcrumb structure of the final gluten-free breads.

Rheological Properties of Gluten-Free Cookies

The values of elastic modulus are presented in Table 6 and can be used for evaluating the influence of BF content on rheological properties of cookie formulations. In all the measurements elastic modulus G' was higher than viscous modulus G" in the whole range of applied frequencies showing minor increase with frequency increase which refers to a solid like behaviour of tested dough samples. It can be observed that increasing the amount of HBF in flour mixtures resulted in higher G' values. The increased viscoelastic properties of the blends containing higher amounts of BF can be assigned to long-chain CMC molecules, containing numerous OH^- groups that participate in strong intermolecular interaction (hydrogen bonding, dipole-dipole, charge effects) with protein molecules. Therefore, blending the long CMC molecules with BF proteins increased molecular entanglements resulting in improved mechanical and viscoelastic properties (Su et al., 2010).

Table 6. Values of storage modulus of gluten-free cookie dough containing rice (RF) and husked buckwheat (HBF) flour [A]

Gluten-free blend	G' (Pa)
70%RF+30%HBF	102698±5395.44[a]
80%RF+20%HBF	84287±2086.97[b]
90%RF+10%HBF	71814±848.53[c]

[A]Values represent the means±standard deviation. Values followed by different lower-case letters are significantly different from each other ($p \leq 0.05$).

Textural and Sensory Analysis of Gluten-Free Cookies

Textural properties of different gluten-free formulations are shown in Table 7. The force required to fracture tested cookie sample can be referred to as the hardness of the sample. It can be observed that cookies with higher content of buckwheat flour expressed higher hardness values. Namely, it has been proved that blending long CMC molecules with different proteins during baking at elevated temperatures resulted in new bonds formation by Maillard reactions that can increase mechanical properties (Ajandouz et al., 2008; Yasir et al., 2007). Also, the increase in mechanical properties may be due to differences in the physicochemical and starch granular properties of rice and buckwheat flour. This could be associated with different swelling behaviour of starch granules of RF and BF having impact on the formation of air zones differing in dimensions. Therefore, this could be influential for creating cookies with different hardness values (Singh et al., 2003).

Table 7. Textural properties of final gluten-free cookies containing rice (RF) and husked buckwheat (HBF) flour [A]

Gluten-free blend	Hardness (g)
70%RF+30%HBF	3251±270[a]
80%RF+20%HBF	2814±181[a]
90%RF+10%HBF	1758±53[b]

[A]Values represent the means±standard deviation. Values followed by different lower-case letters are significantly different from each other ($p \leq 0.05$).

While the increase in BF content has had a remarkable effect on the appearance of gluten-free bread, there was only a minor influence on the appearance of gluten-free cookies which manifested as a change in product colour. Namely, increase in BF content led to product with darker colour (Figure 7).

Figure 7. Appearance of the gluten-free cookies.

However, cookie formulations with 20% and 30% of BF (calculated on flour mass) were more tasteful than formulation containing 10% of BF. Namely, it is well known that rice flour is characterized by bland and neutral taste (Gujral & Rosell, 2004b). Therefore, the increase in the content of BF resulted in more pleasant taste and flavour. Polyphenols, aromatic compound, are commonly found in buckwheat flour and they are responsible for giving the pleasant aroma (Luthar, Z., 1992) masking the bland and neutral rice flour taste.

CONCLUSION

The work described in this chapter and that previously reported (Torbica et al., 2010) suggest that buckwheat enhanced rice flour presents nutritionally valuable base for gluten-free products creation. Except being natural source of minerals and antioxidants, buckwheat flour has also proven as important ingredient in structure formation and sensory improvement.

Although some previously reported papers, as well as scanning electron microscopy analyses conducted in this work, have shown that there is a difference in structure of rice and buckwheat flour to wheat flour, it has been proven that with or without addition of hydrocolloids such as carboxymethilcellulose, blends of rice and buckwheat flour could be used together with other ingredients in preparation of gluten-free bakery products. Namely, according to empirical rheological (Mixolab) measurements as well as electrophoregrams, rice and buckwheat flour proteins have similar properties as the ones of wheat flour, meaning that these proteins are able to form a protein network which can be responsible for structural and mechanical properties of the final product. Except having the ability to develop protein network (factor important in breadmaking), Mixolab profiles revealed that blends of rice and buckwheat flour could mimic wheat starch behaviour during gelatinization (factor important during the manufacture of biscuits and cookies).

Concerning the protein properties the increase in the amount of buckwheat flour in rice/buckwheat flour blend resulted in increase in dough development time and weakening of the protein network and decrease in stability. Concerning starch properties the increase in the amount of buckwheat flour led to decrease in peak viscosity and retrogradation degree.

Gluten-free dough containing unhusked buckwheat flour had higher water absorption values, lower stability and weaker protein network structure than that consisted of husked buckwheat flour, which was the result of higher content of fiber materials present in unhusked buckwheat flour. Moreover, due to lower lipid and protein content, samples containing HBF expressed higher starch swelling capacities than the UBF containing samples which was manifested as higher peak viscosity.

Dynamic shear rheological measurements have shown the possibility of interactions between gluten-free bread dough ingredients, which manifested as an increase in viscoelasticity with the addition of higher content of husked buckwheat flour. The increase in viscoelasticity was more pronounced in cookie dough due to presence of CMC which interacted with other ingredients.

The addition of buckwheat flour did not affect textural properties of gluten-free bread significantly; however, hardness of cookies was increased with the addition of buckwheat flour.

Sensory properties, on contrary, i.e. appearance of gluten-free bread were more affected by buckwheat flour content than the appearance of cookies. The only change in cookie appearance was change in colour, while the addition of buckwheat flour to bread formulations resulted in more cracked upper crust surface.

In general, all gluten-free formulations (bread and cookies) were found sensory acceptable.

REFERENCES

AACC, 2000. Bread firmness by universal testing machine. In *Approved methods of the AACC* (74-09), St. Paul, MN, USA: American Association of Cereals Chemists.

Aubrecht, E. & Biacs, P. Á. (2001). Characterization of buckwheat grain proteins and its products, *Acta Alimentaria*, 28, 261-268.

Ahlborn, G.J., Pike, O. A., Hendrix, S. B., Hess, W. M. & Huber, C. S. (2005). Sensory, mechanical, and microscopic evaluation of staling in low-protein and gluten-free breads. *Cereal Chemistry*, 82, 328–335.

Ajandouz, E. H., Desseaux, V., Tazi, S. & Puigserver, A. (2008). Effects of temperature and pH on the kinetics of caramelisation, protein cross-linking and Maillard reactions in aqueous model systems. *Food Chemistry*, 107(3), 1244–1252.

Alvarez-Jubete, L., Arendt, E. K. & Gallagher, E. (2009). Nutritive value and chemical composition of pseudocereals as gluten-free ingredients. *International Journal of Food Science and Nutrition,* 60(1) (Suppl. 4), 240-257.

Alvarez-Jubete, L., Arendt E. K. & Gallagher, E. (2010). Nutritive value of pseudocereals and their increasing use as functional gluten-free ingredients. *Trends in Food Science & Technology*, 21, 106-113.

Arendt, E. K., Morrissey, A., Moore, M. M. & Bello, F. D. (2008). Gluten-free breads. In E. K. Arendt & F. Dal Bello (Eds.), *Gluten-free cereal products and beverages* (pp. 289 - 311), USA: Elsevier.

Bechtel, D. B. & Wilson, J. D. (2003). Amyloplast formation and starch granule development in hard red winter wheat. *Cereal Chemistry*, 80, 175–83.

Bejosano F. P. & Corke H. (1999). Properties of protein concentrates and hydrolysates from Amaranthus and Buckweat. *Industrial Crops and Products*, 10, 175-183.

Bonafacciaa, G., Marocchinia, M. & Kreft, I. (2003). Composition and technological properties of the flourand bran from common and tartary buckwheat. *Food Chemistry*, 80, 9–15.

Bonet, A., Blaszczak, W. & Rosell, C. M. (2006). Formation of homopolymers and heteropolymers between wheat flour and several protein sources by transglutaminase-catalyzed cross linking. *Cereal Chemistry*, 83(6), 655-662.

Debet, M. R. & Gidley, M. J. (2006). Three classes of starch granule swelling: Influence of surface proteins and lipids. *Carbohydrate Polymers*, 64, 452–465.

Demirkesen, I., Mert, B., Sumnu, G. & Sahin S. (2010). Rheological properties of gluten-free bread formulations. *Journal of Food Engineering*, 96, 295–303

Djaković, Lj. & Dokić, P. (1972). Die Rheologische Characterisierung der Stärkegele. *Die Stärke*, 24, 195-201.

Djaković, Lj. & Dokić, P. (1978). Changes of Viscous Characteristics of Oil in Water Emulsions During Homogenization. *Colloid and Polymer Science*, 256, 1171-1181.

Dobraszczyk, B. J. & Morgenstern, M. P. (2003). Rheology and the breadmaking process. *Journal of Cereal Science*, 38, 229–245.

Dokić, P. & Djaković, Lj. (1975). Rheological Characteristics of β-lipoprotein. *Journal of Colloid and Interface Science*, 51, 373-379.

Dokić, P., Jakovljević, J. & Dokić-Baucal, Lj. (1998). Molecular Characteristics of Maltodextrins and Rheological Behaviour of Diluted and Concentrated Solutions. *Colloids and Surfaces A*, 141, 435-440.

Dokić-Baucal, Lj., Dokić P. & Jakovljević, J. (2004). Influence of different maltodextrins on properties of O/W emulsions. *Food Hydrocolloids*, 18(2), 233-239.

Dokić, P., Dokić, Lj., Dapčević, T. & Krstonošić, V. (2008). Colloid Characteristics and Emulsifying Properties of OSA Starches. *Progress in Colloid and Polymer Science*, 135, 48–56.

Dokić, Lj., Dapčević, T., Krstonošić, V., Dokić, P. & Hadnađev, M. (2010). Rheological characterization of corn starch isolated by alkali method. *Food Hydrocolloids*, 24(2-3), 172–177.

Drzewiecki, J., Delgado-Licon, E., Haruenkit, R., Pawelzik, E., Martin-Belloso, O. & Park, Y.-S. (2003). Identification and differences of total proteins and their soluble fractions in some pseudocereals based on electrophoretic patterns. *Journal of Agricultural and Food Chemistry*, 51(26), 7798-7804.

Fitzsimons, S. M., Mulvihill, D.M. & Morris, E. R. (2008). Large enhancements in thermogelation of whey protein isolate by incorporation of very low concentrations of guar gum. Food Hydrocolloids ,22, 576–586.

Gallagher, E., Kunkel, A., Gormley, R. T. & Arendt, K. E. (2003). The effect of dairy and rice powder addition on loaf and crumb characteristics, and on shelf life (intermediate and long-term) of gluten-free breads stored in a modified atmosphere. *European Food Research and Technology*, 218, 44–48.

Gallagher, E., Gormley, R. T. & Arendt, K. E. (2004). Recent advances in the formulation of gluten-free cereal-based products. *Trends in Food Science & Technology*, 15, 143–152.

Gorinstein, S., Pawelzik, E., Delgado-Licon, E., Haruenkit, R., Weisz, M. & Trakhtenberg, S. (2002). Characterisation of pseudocereal and cereal proteins by protein and amino acid analysis. *Journal of the Science of Food and Agriculture*, 82, 886-891.

Gujral, H. S., Guardiola, I., Carbonell, J. V. & Rosell, C. M. (2003a). Effect of cyclodextrinase on dough rheology and bread quality from rice flour. *Journal of Agricultural and Food Chemistry*, 51, 3814–3818.

Gujral, H. S., Haros, M. & Rosell, C. M. (2003b). Starch hydrolyzing enzymes for retarding the staling of rice bread. *Cereal Chemistry*, 80(6), 750–754.

Gujral, H. S., & Rosell, C. M. (2004a). Functionality of rice flour modified with a microbial transglutaminase. *Journal of Cereal Science*, 39, 225–230.

Gujral, H. S., & Rosell, C. M. (2004b). Improvement of the breadmaking quality of rice flour by glucose oxidase. *Food Research International*, 37(1), 75–81.

Haque, A., & Morris, E. R. (1994). Combined use of ispaghula and HPMC to replace or augment gluten in breadmaking. *Food Research International*, 27, 379–393.

Harrington, J. C., Foegeding, E. A., Mulvihill, D. M. & Morris, E. R. (2009). Segregative interactions and competitive binding of Ca2+ in gelling mixtures of whey protein isolate with Na+ k-carrageenan. *Food Hydrocolloids*, 23, 468–489.

He, J., Klag, J. M., Whelton, P. K., Mo, J.-P., Chen, J.-Y., Qian, M.-C., Mo, P.-S. & He, G.-Q. (1995). Oats and buckwheat intakes and cardiovascular disease risk factors in an ethnic minority of China. *American Journal of Clinical Nutrition*, 61, 366-372.

Holasova, M., Fiedlerova, V., Smrcinova, H., Orsak, M., Lachman, J. & Vavreinova, S. (2002). Buckwheat–the source of antioxidant activity in functional foods. *Food Research International*, 35(23), 207–211.

Hopman, E. G. D., Le Cessie, S., Von Blomberg, M. E. & Mearin, M. L. (2006). Nutritional management of the gluten-free diet in young people with celiac disease in The Netherlands. *Journal of Pediatric Gastroenterology and Nutrition*, 43, 102-108.

Iqbal, A., Khalil, I. A., Ateeq, N. & Khan, M. S. (2006). Nutritional quality of important food legumes, *Food Chemistry*, 97, 331–335.

Janeš, D., Prosen, H., Kreft, I. & Kreft, S. (2010). Aroma Compounds in Buckwheat (*Fagopyrum esculentum* Moench) Groats, Flour, Bran, and Husk. *Cereal Chemistry*, 87(2), 141–143.

Ju, Z.Y., Hettiarachchy, N.S. & Rath, N. (2001). Extraction, denaturation and hydrophobic properties of rice flour proteins. *Journal of Food Science*, 66, 229–232.

Kang, M. Y., Choi, Y. H. & Choi, H. C. (1997). Effects of gums, fats and glutens adding on processing and quality of milled rice bread. *Korean Journal of Food Science and Technology*, 29, 700–704.

Krstonošić, V., Dokić, Lj., Dokić, P. & Dapčević, T. (2009). Effects of xanthan gum on physicochemical properties and stability of corn oilin- water emulsions stabilized by polyoxyethylene (20) sorbitan monooleate. *Food Hydrocolloids*, 23(8), 2212-2218.

Kupper, C. (2005). Dietary guidelines and implementation for celiac disease. *Gastroenterology*,128(4), S121-S127.

Lazaridou, A., Duta, D., Papageorgiou, M., Belc, N. & Biliaderis, C. G. (2007). Effects of hydrocolloids on dough rheology and bread quality parameters in gluten-free formulations. *Journal of Food Engineering*, 79, 1033–1047.

Lin, R., Tao, Y. & Li, X. (1992). Preliminary division of cultural and ecological regions of Chinese buckwheat. Fagopyrum, 12, 48–55

Lin, L.-Y., Liu, H.-M., Yu, Y.-W., Lin, S.-D. & Mau, J.-L. (2009). Quality and antioxidant property of buckwheat enhanced wheat bread. *Food Chemistry*, 112, 987–991.

Lopez, A. C. B., Pereira, A. J. G. & Junqueira, R. G. (2004). Flour mixtures of rice flour, corn and cassava starch in the production of gluten-free white bread. *Brazilian Archives of Biology and Technology,* 47, 63-70.

Luthar, Z. (1992). Polyphenol classification and tannin content of buckwheat seeds (Fagopyrum escelentum Moench). *Fagopyrum,* 12, 36-42.

Marco, C. & Rosell, C. M. (2008a). Breadmaking performance of protein enriched, gluten-free breads. *European Food Research and Technology,* 227, 1205–1213.

Marco, C. & Rosell, C. M. (2008b). Effect of different protein isolates and transglutaminase on rice flour properties. *Journal of Food Engineering,* 84, 132–139.

Marconi, E. & Careca, M. (2001). Pasta from nontraditional raw materials. *Cereal Foods World,* 46, 522–530.

Mariotti, M., Lucisano, M., Pagani, M. A. & Iameti, S. (2008). Macromolecular Interactions and Rheological Properties of Buckwheat-Based Dough Obtained from Differently Procesed Grains. *Journal of Agricultural and Food Chemistry,* 56, 4258-4267.

McCarthy, D. F., Gallagher, E., Gormley, T. R., Schober, T. J. & Arendt, E. K. (2005). Application of response surface methodology in the development of gluten-free bread. *Cereal Chemistry,* 82, 609–615.

Moore, M. M., Schober, T. J., Dockery, P. & Arendt, E. K. (2004). Textural comparisons of gluten-free and wheat-based doughs, batters, and breads. *Cereal Chemistry,* 81, 567–575.

Moore, M. M., Heinbockel, M., Dockery, P., Ulmer, H. M. & Arendt, E. K. (2006). Network formation in gluten-free bread with application of transgluteminase. *Cereal Chemistry,* 83(1), 28–36.

Muller, H. G. (1975). Rheology and the conventional bread and biscuit making process. *Cereal Chemistry,* 52, 89–105.

Murray, J. A. (1999). The widening spectrum of celiac disease. *American Journal of Clinical Nutrition,* 69, 354–365.

Nelles, E. M., Dewar, J., Bason, M. L. & Taylor, J. R. N. (2000). Maize Starch Biphasic Pasting Curves. *Journal of Cereal Science,* 31, 287–294.

Oomah, B. D., & Mazza, G. (1996). Flavonoids and antioxidative activities in buckwheat. *Journal of Agricultural and Food Chemistry,* 44(5), 1746–1750.

Ozturk, S., Kahraman, K., Tiftik, B. & Koksel, H. (2008). Predicting the cookie quality of flours by using Mixolab. *European food research & technology,* 227 (5), 1549-1554.

Prestamo, G., Pedrazuela, A., Penas, E., Lasuncion, A. M. & Arroyo, G. (2003). Role of buckwheat diet on rats as prebiotic and healthy food. *Nutrition Research,* 23, 803–814.

Resurrecion, A.V.A. (2008). Consumer sensory testing for food product development. In: B. A. L. & L. J. B. (Eds.), *Developing New Products for a Changing Marketplace* (second edition, pp. 5-25). USA: CRC Press Taylor and Francis Gorup.

Ribotta, P. D., Ausar, S. F., Morcillo, M. H., Perez, G. T., Beltramo, D. M. & Leon, A. E. (2004). Production of gluten-free bread using soybean flour. *Journal of The Science of Food and Agriculture,* 84, 1969–1974.

Rosell, C. M. & Marco, C. (2008). Rice. In E. A. Arendt & F. Dal Bello (Eds.), *Gluten-free cereal products and beverages.* (pp. 81-100). Oxford, UK: Elsevier.

Schober, T. J., Messerschmidt, M., Bean, S. R., Park, S. H. & Arendt, E. K. (2005). Gluten-free bread from sorghum: quality differences among hybrids. *Cereal Chemistry,* 82, 394–404.

Schoenlechner, R., Siebenhandl, S. & Berghofer E. (2008). Pseudocereals, In E. K. Arendt &F. Dal Bello (Eds.), *Gluten-free cereal products and beverages,* (pp. 289 -311). USA: Elsevier.

Sciarini, S. L., Ribotta, D. P., León, E. A. & Pérez, T. G. (2008). Influence of Gluten-free Flours and their Mixtures on Batter Properties and Bread Quality. *Food and Bioprocess Technology,* 3(4), 577-585.

Singh, J., Singh N., Sharma, T. R. & Saxena, S. K. (2003). Physicochemical, rheological and cookie making properties of corn and potato flours. *Food Chemistry,* 83, 387–393.

Sivaramakrishnan, H. P., Senge, B. & Chattopadhyay, P. K. (2004). Rheological properties of rice dough for making rice bread. *Journal of Food Engineering,* 62, 37–45.

Stauffer, C. E. (2007). Principles of dough formation, In S. P. Cauvain & L. S. Young (Eds.), *Technology of breadmaking,* NY, USA: Springer.

Su, J.-F., Huang, Z., Yuan, X.-Y., Wang, X.-Y. & Li, M. (2010). Structure and properties of carboxymethyl cellulose/soy protein isolate blend edible films crosslinked by Maillard reactions. *Carbohydrate Polymers,* 79, 145–153.

Thompson, T. (1999). Thiamin, riboflavin, and niacin contents of the gluten free diet: is there cause for concern? *Journal of the American Dietetic Association,* 99, 858-862.

Thompson, T. (2000). Folate, iron, and dietary fiber contents of the gluten-free diet. *Journal of the American Dietetic Association,* 100, 1389-1395.

Thompson, T., Dennis, M., Higgins, L. A., Lee, A. R. & Sharrett, M. K. (2005). Gluten-free diet survey: are Americans with coeliac disease consuming recommended amounts of fibre, iron calcium and grain foods? *Journal of Human Nutrition and Dietetics,* 18, 163-169.

Torbica, A., Antov, M., Mastilović, J. & Knežević, D. (2007). The influence of changes in gluten complex structure on technological quality of wheat (Triticum aestivum L.). *Food Research International,* 40, 1038–1045.

Torbica, A., Hadnađev, M. & Dapčević, T. (2010). Rheological, textural and sensory properties of gluten-free bread formulations based on rice and buckwheat flour. *Food Hydrocolloids,* 24, 626-632.

Toufeili, I., Dagher, S., Shadarevian, S., Noureddine, A., Sarakbi, M. & Farran, M. (1994). Formulation of gluten-free pocket-type flat breads: optimization of methylcellulose, gum Arabic, and egg albumen levels by response surface methodology. *Cereal Chemistry,* 71, 594–601.

Veraverbeke, W. S. & Delcour, J. A. (2002). Wheat Protein Composition and Properties of Wheat Glutenin in Relation to Breadmaking Functionality. *Critical Reviews in Food Science and Nutrition,* 42(3), 179-208.

Watanabe, M., Ho, C. T. & Lee, Y. C. (1997). Antioxidant compounds from buckwheat (Fagopyrum esculentum Moench) hulls. *Journal of Agricultural and Food Chemistry,* 45, 1039–1044.

Witczak, M., Korus, J., Ziobro, R. & Juszczak, L. (2010). The effects of maltodextrins on gluten-free dough and quality of bread. *Journal of Food Engineering,* 96, 258–265.

Yasir, S. B. M., Sutton, K. H., Newberry, M. P., Andrews, N. R., Gerrard, J. A. (2007). The impact of Maillard cross-linking on soy proteins and tofu texture. *Food Chemistry,* 104(4), 1502–1508.

Ylimaki, G., Hawrysh, Z. J., Hardin, R. T. & Thomson, A. B. R. (1991). Response surface methodology in the development of rice flour yeast breads: sensory evaluation. *Journal of Food Science,* 56(3), 751–755.

In: Gluten: Properties, Modifications and Dietary Intolerance
Editor: Diane S. Fellstone

ISBN: 978-1-61209-317-8
©2011 Nova Science Publishers, Inc.

Chapter 4

CHIAROSCURO OF STANDARDIZATION FOR GLUTEN-FREE FOODS LABELING AND GLUTEN QUANTITATION METHODS

A. M. Calderón de la Barca, E. J. Esquer-Munguía and F. Cabrera-Chávez

Centro de Investigación en Alimentación y Desarrollo, A.C.
Hermosillo, Sonora, México

ABSTRACT

Celiac disease (CD) is an autoimmune enteropathy characterized by intolerance to wheat gluten. CD patients must adhere to a strict lifelong gluten-free diet, excluding all food products containing wheat and taxonomically related cereals. The CD prevalence is 1% in any population over the world and apparently it is increasing. Thus, the gluten content must be regulated in specialties for gluten intolerant patients by reliable and sensitive methods. To reach this objective, several procedures have been used including immunological and non-immunological methods. Additionally to limitations on gluten quantitation due to the principles of the tests, there are limitations due to the methods of proteins' extraction. In this chapter, troubles and trends for gluten quantitation in gluten-free foodstuffs are discussed. Also differences and agreements on the regulations of gluten-free labeling as well as the related basic definitions are shown. Finally, it is stated the consideration of the industry needs, current scientist knowledge and the safety of gluten intolerant patients for a good regulation and international trade of gluten-free foodstuffs.

INTRODUCTION

Since wheat was domesticated in the Middle East and spread to Europe by about 5,000 years ago, it has become the base of the occidental diet. However, for some individuals the wheat proteins could be deleterious, such as for celiac disease (CD) or gluten intolerant patients. CD is a lifelong inflammatory disease of the small intestine caused by exposition to dietary gluten in genetically predisposed individuals. Its pathogenesis is mediated by IgA antibodies to gluten proteins of wheat, rye and barley as well as to the tissue transglutaminase (tTG), the autoantigen (Di Sabatino and Corazza, 2009). Nowadays CD is recognized to be widely distributed with an estimated prevalence of 1% in the general population (Schuppan et al., 2009), including non-Caucasians of some regions and countries where few years ago it was considered to be rare (Catassi and Cobellis, 2007).

The CD treatment involves the removal of all dietary gluten, which helps to improve health and nutritional status of celiac patients and reduces the risk of other complications (Yachha et al., 2007; Annibale et al., 2001). Therefore, the industry in industrialized countries has commercialized a wide variety of gluten-free foodstuffs to satisfy the demand, since many years ago. In addition to the *Codex Alimentarius* standard for gluten-free labeling of foods, several countries present their regulation and monitoring system. Because the new map distribution of CD, new gluten-free products available and appropriate to each regional taste are needed, as well as the respective legislation.

The *Codex Alimentarius* standard applies to foods for special dietary uses that have been formulated, processed or prepared to meet the special dietary needs of people intolerant to gluten (*Codex Alimentarious*, 2008). Also, the *Codex* recommends gluten quantitation by the enzyme-linked immunosorbent assay (ELISA), using monoclonal antibodies against the 5-residues gluten peptide (R5). In the present chapter some of the problems for gluten quantitation are discussed, as well as the limitations of the used techniques for its evaluation.

GLUTEN PROTEINS AND CELIAC DISEASE

Wheat gluten proteins are a major component of wheat flour and are responsible for the ability of flour to form dough with the right rheological properties for baking. Although the term gluten strictly encompasses proteins in wheat, for simplicity all the related proteins from other cereals as rye and barley, which exacerbate CD, are called gluten proteins (Kagnoff, 2007). Gluten includes two principal protein types, the glutenins and gliadins. In the current draft standard of the *Codex Alimentarius*, gliadins are defined as the prolamins from wheat, a gluten fraction extractable by 40-70% of ethanol, although part of the glutenins are also soluble in alcohol after reduction (Bean et al., 1998). Apparently the trend is to use prolamins instead of gluten proteins, but currently the prolamin content of gluten is taken as 50%. Additionally, prolamins from rye are secalins, from barley hordein and from oats avenins.

Prolamins contain proline- and glutamine-rich sequences that are resistant to complete proteolytic digestion by the gastrointestinal peptidases and therefore, peptides as long as the 33-mer of α-gliadins is still undigested and prone to activate the CD immune response (Rodrigo, 2009). The long gluten peptides alone are not sufficient to cause CD, the rich-glutamine and proline peptides are bound to receptor molecules HLA-DQ2 or DQ8 (genetically determined) in the antigen presenting cells and therefore, to the T-cells to start the immune response (Kagnoff, 2007).

For regulation, the *Codex Alimentarius* defines gluten as "a protein fraction from wheat, rye, barley, oats or their cross varieties and derivatives thereof, to which some people are intolerant and that is insoluble in water and 0.5 M NaCl" (*Codex Alimentarius*, 2008). On the same way, according to the US Food and Drug Administration (FDA) gluten are "the proteins that naturally occur in a prohibited grain and that may cause adverse health effects in persons with celiac disease". "Prohibited grains" are wheat, rye, barley or a crossbred hybrid of these grains (FDA, 2007).

Gluten definition based on solubility could be not adequate for CD patients' safety, because food processing could increase protein solubility but not necessarily reduce its adverse effects. Instead, the FDA's definition allows the consideration of peptides from gluten involved in CD triggering, although it does not take into account other proteins that may cause adverse health effects in some CD sufferers. For example, maize prolamins induce an immune response in some CD cases (Kristjansson et al., 2005; Cabrera-Chavez et al., 2008).

SOURCES OF GLUTEN-FREE FOODSTUFFS CROSS-CONTAMINATION

Naturally gluten-free foodstuffs lose their status due to cross-contamination with wheat gluten. It includes the contamination in production cycle, harvesting, transportation and storage. In the milling environment and during processing contamination occurs if the same equipment is used for the production, processing and packaging of gluten-containing and gluten-free products. At the primary production level, gluten-free grains may become cross-contaminated by gluten-containing cereals unless preventative measures are taken for harvesting, transportation and grains' storage. At processing, gluten-free products are susceptible to cross-contamination by machinery and equipments as well as by gluten containing flour dust in the air or through the hands and clothes of workers. The risk of gluten contamination increases with the number of processing steps (Størsrud et al., 2003). Also during the transport and storage of gluten-free foodstuffs there are possibilities of contamination by mixing up with gluten-containing products.

GLUTEN AS ADDITIVE IN SOME NON-WHEAT FOODSTUFFS

Some naturally gluten-free based foodstuffs could contain wheat gluten intentionally added as "vital gluten" for improving functional properties. Vital gluten is dried wheat gluten obtained by the washing of wheat dough and drying at different temperatures (Day et al., 2006). Although it was first used as a dough strengthener, formulation and processing aids for baking, currently it is added as stabilizer, thickener and texturing agent for several foodstuffs. As a matter of fact, ~13% of the vital gluten commercialized around the world is used for non-wheat products (Day et al., 2006).

Some of the food products with added gluten could include processed hamburger and other meat and fish products, sauces, sausages and baby foods (Astuta et al., 1999; Hand et al., 1981; Blandino et al., 2003). In most of the industrialized countries, the labeling of this kind of product is not a problem for CD or gluten intolerant patients because the governments regulate it. However, in emerging countries the label laws frequently do not exist.

REGULATION OF LABELING OF GLUTEN-FREE PRODUCTS

The gluten-free labeling should be regulated in foods according to the quantity of gluten that can be tolerated by CD patients. By 1981 the *Codex Alimentarius* had established 500 mg/kg (ppm) as the upper limit for gluten content in gluten-free products. However, there were reports of negative effects on CD patients after eating foods with less than 500 ppm (Biagi et al., 2001; Hischenhuber et al., 2006). Therefore, the limit was reduced to less than 200 ppm gluten but based on results of well controlled trials establishing than CD patients tolerate 10 ppm of gluten per day, whereas 50 ppm is harmful (Catassi et al., 2007), the standard has changed.

The current revised standard of the *Codex Alimentarius* defines that gluten-free foods are dietary foods: a) consisting of or made only from one or more ingredients which do not contain wheat (i.e., all *Triticum* species, such as durum wheat, spelt, and kamut), rye, barley, oats or their crossbred varieties, and the gluten level does not exceed 20 ppm in total, based on the food as sold or distributed to the consumer, and/or b) consisting of one or more ingredients from wheat (i.e., all *Triticum* species, such as durum wheat, spelt, and kamut), rye, barley, oats or their crossbred varieties, which have been specially processed to remove gluten, and the gluten level does not exceed 20 ppm in total, based on the food as sold or distributed to the consumer (*Codex Alimentarius*, 2008).

Additionally, the *Codex Alimentarius* defines foods specially processed to reduce gluten content as these foods consisting of one or more ingredients from wheat (i.e., all *Triticum* species, such as durum wheat, spelt, and kamut), rye, barley, oats or their crossbred varieties, which have been specially processed to reduce the gluten content to a level above 20 up to 100 ppm in total, based on the food as sold or distributed to the consumer. Decisions on the marketing of products described in this section may be determined at the national level.

In August 2004, the US Food Allergen Labeling and Consumer Protection Act (FALCPA) of 2004 became a law. This legislation required manufacturers to identify the eight major food allergens, including wheat (but not barley and rye) on the food label; it became effective on January 1, 2006. The FALCPA also mandated the FDA to issue a proposed rule to define and permit the use of the term "gluten-free" on food labels by August 2006, with the final ruling by August 2008. The proposed gluten-free regulation was released on January 2007 and the FDA reviewed comments from consumers, industry, health professionals and others. The final rule to establish a regulatory definition for the term "gluten-free" was expected in August 2008; however it has been delayed until a safety assessment report on gluten exposure in individuals with celiac disease has been published (Case, 2010).

The term that the US FDA is proposing to define the food-labeling term gluten-free is based on the voluntary use in the labeling of foods, to indicate that the food does not contain any of the following: An ingredient that is any species of the grains wheat, rye, barley, or a crossbred hybrid of these grains (all noted grains are collectively referred to as "prohibited grains"); an ingredient that is derived from a prohibited grain and that has not been processed to remove gluten (e.g., wheat flour); an ingredient that is derived from a prohibited grain and that has been processed to remove gluten (e.g., wheat starch), if the use of that ingredient results in the presence of 20 ppm or more gluten in the food; or 20 ppm or more gluten.

A food that bears the claim "gluten-free" or similar claim in its labeling and fails to meet the conditions specified in the proposed definition of "gluten-free" would be deemed misbranded. FDA also is proposing to deem misbranded a food bearing a gluten-free claim in its labeling if the food is inherently free of gluten and if the claim does not refer to all foods of that same type (e.g., milk, a gluten-free food or all milk is gluten-free). In addition, a food made from oats that bears a gluten-free claim in its labeling would be deemed misbranded if the claim suggests that all such foods are gluten-free or if 20 ppm or more gluten is present in the food.

There are still differences regarding the regulation and definition of gluten-free products around the world (Niewinski, 2008). It causes confusion for CD patients and the food industry, resulting in various interpretations of gluten-free and labeling (Case, 2010). The majority of the European countries have accepted the definition of gluten-free labeling by the *Codex Alimentarius*. On the other hand, countries like New Zealand and Australia established that in order to be labeled gluten-free, a food must contain "No detectable gluten" by the most sensitive universally accepted test method. At the time of the printing of the Ingredient List, 7th Edition, testing achieved a detection level of 0.0005 (5 ppm). If gluten is not detected then the food can be labeled gluten-free (Koeller and La France, 2007).

Health Canada proposed a new regulation on July 26, 2008 entitled Enhanced Labelling of Food Allergen and Gluten Sources and Added Sulphites which will require manufacturers to declare on the food label the major food allergens, all gluten sources and sulphites when present as ingredients or components of ingredients. Until the final mandatory amendments become law, Health Canada and the Canadian Food Inspection Agency strongly urge manufacturers to declare on their food labels the allergens, gluten sources and sulphites. Canada has a specific regulation for the term "gluten-free" that was established over 25 years ago. On May 13, 2010 Health Canada announced that the gluten free regulation (B.24.018) was under review, including the labeling of pure, uncontaminated oats.

Regardless of these regulations, in some parts of the world mainly in emerging countries, there is still a lack of information and norms about gluten-free labeling, exposing CD or gluten intolerant patients to have relapses and a slow recovery from the disease.

METHODS FOR GLUTEN QUANTITATION

Different methods of analysis are used to assess the gluten content of food. According to the *Codex Alimentarius* standard for gluten-free food the analytical method for testing different foods should be based on an immunochemical determination of gliadins. Homemade or commercial enzyme-linked immunosorbent assay (ELISA) methods using different monoclonal or polyclonal antibodies against a variety of gliadins or their peptides are commonly used. Besides the immunological methods, non-immunologic procedures like mass spectrometry, high performance liquid chromatography (HPLC) or molecular methods like PCR, have demonstrated to be efficacious and powerful in the analysis of wheat gliadins, barley hordeins, rye secalins, and oat avenins, becoming more and more applicable (Méndez et al., 2000).

IMMUNOLOGICAL METHODS

Immunochemistry is the base of methods for detecting food contamination since it provides specific and sensitive recognition of the contaminant. Also immunochemical tests are less expensive than other techniques and provide rapid results (Denery-Papini et al., 1999).

The ELISA (Enzyme-linked Immunoassay) detects or measures the concentration of the protein of interest in a sample that may contain many different proteins. In the ELISA "sandwich", antibodies are bound to a support plate for specifically binding to the target proteins (which function as antigens) in a solution mixture. Thereafter, incubation is done with second antibodies conjugated to an enzyme, which produce a color in the solution by adding a substrate. Quantitation is done by interpolation in a standard curve of the protein of interest (Méndez et al., 2000).

There are commercially available ELISA kits for gluten measurement in foodstuffs. The ELISA ω-gliadin is the official method of analysis for gluten approved by the AOAC in 1991. This assay uses monoclonal antibodies to ω-gliadins. Due to the antibody specificity against "heat-stable" ω-gliadins, the assay provides accurate results with heat-treated food products. However, the method bears a considerable source of systematic errors considering the differences in the proportions of ω-gliadins in cultivars (Seilmeier and Wiser, 2003) and is unable to detect small gluten fragments and barley sequences (Lester, 2008).

The most accepted immunological method is based on monoclonal antibodies raised against a peptide of rye secalins. In 2006, the *Codex* Committee on Methods of Analysis and Sampling endorsed the sandwich R5 ELISA developed by Méndez (Valdés et al., 2003; Méndez et al., 2005) as a method for determination of gluten content in gluten-free foods. This assay is directed against an amino acid sequence homologous to those assumed to be associated with gluten toxicity. The potentially toxic epitope in celiac disease has been recognized as QQPFP (glutamine-glutamine-proline-phenylalanine-proline, R5), which occurs repeatedly in α-, γ- and ω-gliadins, hordeins and secalins, of wheat, barley and rye to a similar degree (Konic-Ristic et al., 2009). In addition, it also recognizes the potentially toxic 33-mer for CD patients. This method has a detection limit of 3 ppm and a quantitation limit of 5 ppm (Thompson and Méndez, 2008).

The gluten in some food products as malt, beer, syrups and baby foods, is fully or partially hydrolyzed. As a result some of the gluten peptides may contain a single epitope QQPFP (Thompson and Méndez, 2008) and, therefore cannot be detected by the ELISA-R5 sandwich, resulting in underestimates of the levels of prolamins (Valdés et al., 2003). To solve this particular issue a competitive ELISA based on the R5 monoclonal antibodies has been developed. This competitive assay has a detection level of 3 ppm of gluten (Ferré et al., 2003).

Another immunological technique frequently used as a confirmatory method in the detection of gluten in foods is the western blot R5 (Gil et al., 2003). This method consists of a first separation of proteins, according to their molecular weight, by electrophoresis in SDS-polyacrylamide gels. Thereafter, proteins in gels are electro-transferred to PVDF membranes, and membranes incubated with R5-antibodies conjugated to peroxidase. Enzyme activity is developed to produce chemiluminescence for a fine detection.

NON IMMUNOLOGICAL METHODS

Non-immunological methods have been applied to cover and complement some of the flaws that ELISA's technique shows; like the occasional occurrence of false-positive results (Méndez et al., 2000). These also have been used as confirmatory techniques. Some of these newly exploited methods in gluten detection are the polymerase chain reaction (PCR), mass spectrometry (MALDI-TOF MS) and high performance liquid chromatography (HPLC).

The polymerase chain reaction (PCR), a DNA-based method of high specificity and sensitivity, could be an effective alternative for the detection of wheat or other gluten-containing cereals (Olexová et al., 2006). The detection of gluten DNA in a food sample frequently implies that the sample must contain gluten. Therefore, specific primers are designed to amplify gluten proteins' DNA sequences by DNA polymerase. The amplified product is visualized by staining with a fluorescent dye after agarose gel electrophoresis (Pereira da Silva et al., 2010). However conventional PCR does not quantify the amount of DNA in the sample. For quantitative determinations, a Real Time PCR has been developed (IMBIOSIS, Technical Dossier).

Another non-epitope dependent system in the detection of gluten is the Matrix-assisted laser desorption ionization time of flightmass spectrometry (MALDI-TOF MS). This technique has demonstrated to be efficacious and powerful in the analysis of wheat gliadins, barley hordein, rye secalin, and oat avenins. This non-immunological method was the first one of this category used to quantify gluten gliadins in food samples (Camafeita et al., 1998). One of the advantages of MALDI-TOF MS over conventional techniques is the possibility of analyzing protein extracts within a few minutes, without fragmentation, using only small amounts of a sample (Aagaard et al., 2002). The method is based on the direct observation of the prolamins patterns.

Mass spectra of wheat, barley and rye prolamins show they have different patterns: gliadins have molecular masses between 30 and 55 kDa, with characteristic peaks between 30-35 kDa; hordeins have molecular masses between 30 and 45 kDa; and secalins have two characteristic peaks at 32 and 39 kDa. This methodology can be applied to characterize glutenins (Ferranti et al., 2007). Therefore MALDI-TOF-MS can be preliminarily established as a unique system with the ability to discriminate the specific type of gluten toxic fractions present in food samples.

HPLC and gel electrophoresis have been applied to quantify gluten. HPLC allows the separation and qualitative and quantitative determination of compounds of analytical interest. A widely used HPLC technique is reversed-phase high performance liquid chromatography (RP-HPLC). In reversed-phase systems, the stationary phase is slightly polar or nonpolar, while the mobile phase has stronger polarity. An RP-HPLC system has already been described for the separation and quantitative determination of wheat prolamins in foods (Pereira da Silva et al., 2010). Size-exclusion (SE) HPLC has been used to relate the quantity of gluten protein fractions to flour properties. Because of the high separation efficiency, these methods can give the proportions of all gluten protein types.

CHIAROSCURO OF METHODS FOR GLUTEN QUANTIFICATION

The gluten quantification methods have their strengths and limitations, which are important to know in order to obtain better assay results (Table 1). ELISA procedures' limitations consisted mainly in the fact that these are epitope dependent methods. An important issue is the fact that gluten precise composition varies between species and cereal variety. In addition, the food industry processing for the preparation of ingredients and foodstuffs modifies gluten in several ways. The modifications by processing as well as by matrix effects, may affect gluten solubility, its molecular associations, not to mention the availability and the sequence of its epitopes (Lester, 2008).

The ELISA by Skerrit and Hill (1990) that is based on the detection of ω-gliadins by monoclonal antibodies presents the variation of cereal composition between species as its major drawback. The ω-gliadins represent only a low percentage of total gliadins, and their concentration varies from 6 to 20% of total prolamins from wheat (Denery-Papini et al., 1999). Thus there is a risk of over or underestimation of total gliadins and therefore gluten, depending on the composition of a product relative to the control standard. Also the ELISA ω-gliadins are unable to accurately detect and quantify barley prolamins and to accurately quantify hydrolyzed gluten (Pereira da Silva et al., 2010).

Moreover, the R5 sandwich ELISA also has some limitations. It has been described that the R5 sandwich ELISA may overestimate barley prolamins' content in barley contaminated foods, when gliadins standard is used; it is recommended to use a hordein standard (Hernando et al., 2008). Also like the ω-gliadins ELISA, the R5 cannot quantify hydrolyzed gluten. However the R5 competitive ELISA was developed to solve this particular issue with hydrolyzed foods.

One of the considerations to be taken into account for gluten quantitation is whether or not to multiply by the factor 2. It has been shown in some studies (Wieser and Koehler, 2009) that calculation of gluten content by multiplying the prolamin content by factor 2 is not valid. There may be an overestimation in the gluten content when gluten-free products contaminated with wheat, rye, barley and oats are analyzed. Also in the case of wheat starch, the gluten content can be over or under-estimated when multiplying by 2.

An important aspect to consider for gluten quantitation is the effect of the processing and heating of analyzing foods. One of the critical steps in the gluten quantification is the gluten proteins' extraction, which is affected by processing as dry or humid thermal treatments, pressure, and by the food matrix composition than can modify the proteins' structure. The structural changes and the heterogeneity of the food matrices create barriers and negatively affect the gluten extraction from foods (Poms and Anklam, 2004).

Garcia et al. (2005), created the cocktail extraction solution, which is based on reducing and denaturing agents (2-mercaptoethanol and guanidine hydrochloride) that dissolve aggregates of gluten proteins in processed foodstuffs. The cocktail solution helps to open the conformation of gluten proteins and promotes their extraction in 60% ethanol, allowing proteins to be fully extracted and then analyzed by the R5 sandwich ELISA. However, 2-mercaptoethanol is associated with some disadvantages due to the weak reducing power, the toxicity and the unpleasant aroma (Gessendorfer et al., 2010).

Immunological methods are not the only ones that presented limitations for gluten quantitation. Molecular procedures like PCR have certain flaws as fragmentation of DNA during food processing. Additionally, PCR is an indirect method since it does not quantify the presence of gluten, but the DNA that codes for it (Zeltner et al., 2009). Limitation of the mass spectrometry MALDI-TOF for gluten quantitation, lies in the overlapping of mass signals when analysis of gliadins or avenins from foods, which were developed with cereal mixtures are made (Méndez et al., 2000). Thus, it is essential to take into account the limitations of the different methods used in the detection and quantification of gluten in foods, since the degree of reliability in the product labeling will depend on the method used for analysis and the foodstuffs analyzed (Table 1).

Table I. Limitations of immunological (ELISA, western blot) and non-immunological (PCR, MALDI-TOF, HPLC) methods for detection of gluten.

Type of test	Sustentation	Drawbacks
ω-gliadin ELISA	Monoclonal antibodies to ω-gliadin.	Underestimates barley prolamin content in barley-contaminated foods. Cannot quantify gluten that has been hydrolyzed. May over- or underestimate gluten content depending on gliadin standard used.
Sandwich R5 ELISA	R5 monoclonal antibodies to epitope QQPFP, present in all fractions of wheat gliadin, barely hordein and rye secalin.	Overestimates barley prolamin content in barley-contaminated foods (when Prolamin Working Group gliadin standard is used). Cannot quantify gluten that has been hydrolyzed.
Competitive R5 ELISA	R5 monoclonal antibodies to epitope QQPFP.	Only compatible with an ethanol extraction. Cannot accurately assess the gluten content of food containing both heated and hydrolyzed gluten.
Western Blot R5	R5 monoclonal antibodies to epitope QQPFP.	Semi-quantitative method
Real Time PCR	DNA that codes for gluten.	Indirect method. Fragmentation of DNA during food processing.
Mass spectrometry MALDI-TOF	Mass spectra of prolamins from wheat, barley and rye.	Overlapping of mass signals when analysis of gliadins or avenins from foods made of cereals mixtures.
High Performance Liquid Chromatography (HPLC)	The separation of prolamins of analytical interest.	Requires long time and work in samples preparation.

TRENDS IN GLUTEN CONTENT MEASUREMENT

Some of the limitations in the quantitation of gluten listed above have encouraged the development of alternatives and solutions for more accurate analysis, to provide greater safety to the CD or gluten intolerant patients. One of these alternatives is the use of different standards depending on the nature of the test sample in the R5 ELISA. For example, better results will be obtained if a hordein standard were used in foods containing mainly barley; this works in the same way for the other prolamins.

Some considerations have been taken in respect to multiplying by factor 2 for calculation of total gluten from gliadins evaluation. Currently it has been clarified that if a food product is contaminated with barley hordeins, only gluten content can be overestimated unless either a hordein standard is used or the gliadin standard is not multiplied by 2 (Wieser and Koehler, 2009).

To avoid the problems with the use of mercaptoethanol in the cocktail extraction solution for the R5 ELISA quantitation, Gessendorfer et al (2010) replaced it with Tris (2-carboxyethyl) phosphine. This reduction agent in combination with guanidine extracts prolamins from foods treated by heating as effectively as the original cocktail solution. The phosphine-compound has the advantages of low toxicity and no strong odors, as well as allowing the use of low concentrations due to its high reducing power.

It is important to direct further investigations toward new strategies to address the main constraints, and to be constantly improving methods of detection and quantitation of gluten. In the same way, the methods should try to find better international agreements to regularize and achieve global standardization.

Conclusion

The analysis of gluten is challenging because it is a mixture of water-insoluble proteins derived from wheat, barley or rye grains. In manufactured foods gluten proteins are within a wide range of matrices in addition to the protein structural modifications resulting from heating and processing. The commercial R5 ELISA has been deemed sufficiently reliable and sensitive to support standards for gluten-free foods based on final gluten content. If the method is to be widely employed, its limitations should be recognized to use it correctly. For a better international agreement in regulation and standardization, definitions of gluten proteins, allowed gluten content in gluten-free foods, and other basic concepts need to be concordant among industry, scientists, and most important the safety of gluten intolerant patients.

Acknowledgment

This Chapter was written as part of a project financially supported by the Mexican Council for Science and Technology (CONACyT) through the Grant CB-2008/106227. The authors are grateful to CONACyT for fellowships given to Esquer-Munguía and Cabrera-Chávez for M.Sci. and Ph.D. studies, respectively.

REFERENCES

Aagard, H; Sperotto, MM; Petersen, M; Kesmir, C; Radzikowski, L; Jacobsen, S; Sondergaard, I. Variety identification of wheat using mass spectrometry with neutral networks and the influence of mass spectra processing prior to neural network analysis. *Rapid Communications in Mass Spectrometry*, 2002, 16, 1232-1237.

Annibale, B; Severi, C; Chistolini, A; Antonelli, G; Lahner, E; Marcheggiano, A; Iannoni, C; Monarca, B; Delle, FG. Efficacy of gluten-free diet alone on recovery from iron deficiency anemia in adult celiac patients. *The American Journal of Gastroenterology*, 2001, 96,132-7.

Atsuta, E; Maeda, M; Karoru, S; Kawanari, M. *Method of preparing low fat sausage*, 1999, United States Patent No. 5,948,462.

Biagi, F; Campanella, J; Martucci, S; Pezzimenti, D; Ciclitira, PJ; Ellis, HJ; Corazza, GR. A milligram of gluten a day keeps the mucosal recovery away: A case report. *Nutrition Reviews*, 2001, 62, 360-363.

Bean, SR; Lyne, RK; Tilley, KA; Chung, OK; Lookhart, GL. A Rapid *Method for Quantitation of Insoluble Polymeric Proteins in Flour1 Cereal Chem*. 1998, 75, 374–379.

Blandino, A; Al-Aseeri, ME; Pandiella, SS; Cantero, D; Webb, C. Cereal-based fermented foods and beverages. *Food Research International*, 2003, 36, 527–543.

Cabrera-Chávez, F; Rouzaud-Sández, O; Sotelo-Cruz, N; Calderón de la Barca, AM. Transglutaminase treatment of wheat and maize prolamins of bread increases the serum IgA reactivity of celiac disease patients. *Journal of Agricultural and Food Chemistry*, 2008, 56, 1387-1391.

Camafeita, E; Solís, J; Alfonso, P; López, JA; Sorell, L; Méndez, E. Selective identification by matrix-assisted laser desorption/ionization time-of-flight mass spectrometry of different types of gluten in foods made with cereal mixtures. *Journal of Chromatography* 1998, 823, 299-306.

Case, S. *Celiac Disease and the Gluten Free Diet. Gluten-Free Diet: A Comprehensive Resource Guide*. Regina, Saskatchewan, Canada:Case Nutrition Consulting SK; 2010.

Catassi, C; Cobellis, G. Coeliac disease epidemiology is alive and kicking, especially in the developing world. *Digestive and Liver Disease*, 2007, 39, 908-10.

Catassi, C; Fabiani, E; Iacono, G; D'Agate, C; Francavilla, R; Biagi, F; Volta, U; Accomando, S; Picarelli, A; De Vitis, I; Pianelli, G; Gesuita, R; Carle, F; Mandolesi, A; Bearzi, I; Fasano, A. A prospective, double-blind, placebo-controlled trial to establish a safe gluten threshold for patients with celiac disease. *American Journal of Clinical Nutrition*, 2007, 85,160-6

Codex Alimentarius Commission. Draft revised Standard for gluten-free foods [online]. 2008 [2010-08-15] Available from: URL: *http://www.codexalimentarius.net*.

Day, L; Augustin, MA; Batey, IL; Wrigley, CW. Wheat-gluten uses and industry needs. *Trends in Food Science and Technology*, 2006, 17, 82–90.

Denery-Papini s, S; Nicolas Y; Popineau Y. Efficiency and limitations of immunochemical assays for the testing of gluten-free foods. *Journal of Cereal Science*, 1999, 30, 121-131.

Di Sabatino, A; Corazza, GR. Coeliac disease. *Lancet*, 2009, 373,1480-1493.

FDA. US Department of Health and Human Services; Food and Drug Administration: Food labeling; Gluten-free labeling of foods. Proposed rule. Fed Reg 2007, 72: 2795-2817. [online]. 2007 [2010-08-20] Available from: URL: *http://www.cfsan.fda.gov*

Ferranti, P; Mamone, G; Picarello, G; Addeo, F. Mass spectrometry analysis of gliadins in celiac disease. *Journal of Mass Spectrometry*, 2007, 42, 1531–1548.

Ferré, S; García, E; Méndez, E. *Measurment of hydrolysed gliadins by a competitive ELISA based on monoclonal antibody R5: Analysis of syrups and beers.* In: Stern M, editor. Proceedings of the 18th Meeting Working Group on Prolamin Analysis and Toxicity, 2-5 October, 2003, Stockholm, Sweden. Zwickau: Verlag Wissenschaftliche Scripten; 2004:65-69.

Garcia, E; Llorente, M; Hernando, A; Kieffer, R; Wieser, H; Méndez, E. Development of a general procedure for complete extraction of gliadins for heat processed and unheated foods. *European Journal of Gastroenterology & Hepatology*, 2005, 17, 529-539.

Gessendorfer, B; Wieser, H; Koehler, P. Optimisation of a solvent for the complete extraction of prolaminas form heated foods. *Journal of Cereal Science*, 2010, article in press.

Gil, JR; Sbihi, Y; Alvarez, A; Maache, M; Larrubia, M; Rojas, J; Osuna, A. Development of a Dot-Blot System to Detect Gluten in Food. *Food and Agricultural Immunology*, 2003, 15, 235-242.

Hand, L; Crenwelge, CH; Terrell, RN. Effects of wheat gluten, soy isolate and flavorings on properties of restructured beef steaks. *Journal of Food Science*, 1981, 46, 1004-1006.

Hernando, A; Mujico, J; Mena, MC; Lombardía, M; Méndez, E. Measurement of wheat gluten and barley hordeins in contaminated oats from Europe, the United States and Canada by Sandwich R5 ELISA. *European Journal of Gastroenterology & Hepatology*. 2008, 20, 545-554.

Hischenhuber, C; Crevel, R; Jarry, B; Mäki, M; Moneret-Vautrin, DA; Romano, A; Troncone, R; Ward, R. Review article: safe amounts of gluten for patients with wheat allergy or coeliac disease. *Alimentary Pharmacology & Therapeutics*, 2006, 23,559-75.

Kagnoff, MF. Celiac disease: pathogenesis of a model immunogenetic disease. *The Journal of Clinical Investigation*, 2007, 117, 41–49.

Koeller K, La France R. Gluten-Free Labeling Regulations Worldwide: Not All Definitions are Equal! [online]. 2007 [2010-08-21]. Available from: URL: *http://www.glutenfreeda.com/mar07_eating-out-gf.asp*

Konic-Ristic, A; Dodig, D; Krstic, R; Jelic, S; Stankovic, I; Ninkovic, A; Radic, J; Besu, I; Bonaci-Nikolic, B; Jojic, N; Djordjevic, M; Popovic, D; Juranic, Z. Different levels of humoral immunoreactivity to different wheat cultivars gliadin are present in patients with celiac disease and in patients with multiple myeloma. *BMC Immunology*, 2009,10, doi: 10.1186/1471-2172-10-32.

Kristjansson, G; Högman, M; Venge, P; Hällgren, R. Gut mucosal granulocyte activation precedes nitric oxide production: studies in coeliac patients challenged with gluten and corn. *Gut*, 2005,54, 769-774.

Lester, D. Gluten measurement and its relationship to food toxicity for celiac disease patients. *Plant Methods*, 2008,4, doi:10.1186/1746-4811-4-26.

Méndez, E; Valdés, I; Camafeita, E. Analysis of Gluten in Foods by MALDI-TOFMS. *Methods in Molecular Biology*. 2000,146, 355-367.

Méndez, E; Vela, C; Immer, U; Janssen, FW. Report of a collaborative trial to investigate the performance of the R5 enzyme linked immunoassay to determine gliadin in gluten-free food. *European Journal of Gastroenterology & Hepatology*, 2005,17, 1053-63.

Niewinski, MM. Advances in Celiac Disease and Gluten-Free Diet. *Journal of the American Dietetic Association*, 2008,108, 661-672.

Olexová, L; Dovicovicová, LA; Svec, M; Siekel, P; Kuchta, T. Detection of gluten-containing cereals in flours and "gluten-free" bakery products by polymerase chain reaction. *Food Control*, 2006, 17, 234–237.

Pereira da Silva, MM; González-Garcia, MB; Antonius, HP; Delerue-Matos Santos-Silva, A; Costa-García, A. Celiac disease diagnosis and gluten-free food analytical control. *Analytical and Bioanalytical Chemistry*, 2010, 397,1743–1753.

Poms, R; Anklam, E. Effects of Chemical, Physical, and Technological Processes on the Nature of Food Allergens. *Journal of AOAC International*, 2004, 87, 1466-1474.

Rodrigo L. Investigational therapies for celiac disease. *Expert Opinion on Investigational Drugs*, 2009, 12, 1865-1873.

Schuppan, D; Junker, Y; Barisani, D. Celiac disease: from pathogenesis to novel therapies. *Gastroenterology*, 2009, 137, 1912-33.

Seilmeier, W; Wieser, H. Comparative investigations of gluten proteins from different wheat species. IV. Reactivity of gliadin fractions and components from different wheat species in a commercial immunoassay. *European Food Research and Technology*, 2003, 217, 360–364.

Skerrit, JH; Hill, AS. Monoclonal antibody sandwich enzyme immunoassays for determination of gluten in foods. *Journal of Agricultural and Food Chemistry*, 1990, 38, 1771-1778.

Storsrud, S; Malmheden, I; Lenner, RA. Gluten contamination in oat products and products naturally free from gluten. *European Food Research and Technology*, 2003, 217, 481–485.

Thompson, T; Mendez E. Commercial assays to assess gluten content of gluten-free foods: why they are not created equal. *Journal of the American Dietetic Association*, 2008, 108, 1682-7.

Valdés, I; García, E; Llorente, M; Méndez, E. Innovative approach to low-level gluten determination in foods using a novel sandwich enzyme-linked immunosorbent assay protocol. *European Journal of Gastroenterology & Hepatology*, 2003 15, 465-74.

Wieser, H; Koehler, P. Is the calculation of the gluten content by multiplying the prolamin content by a factor of 2 valid? *European Food Research and Technology*, 2009, 229, 9–13.

Yachha, SK; Srivastava, A; Mohindra, S; Krishnani, N; Aggarwal, R; Saxena, A. Effect of a gluten-free diet on growth and small-bowel histology in children with celiac disease in India. *Journal of Gastroenterology and Hepatology*, 2007, 22,1300-5.

Zeltner, D; Glomb, MA; Maede, D. Real-time PCR systems for the detection of the gluten-containing cereals wheat, spelt, kamut, rye, barley and oat. *European Food Research and Technology*, 2009, 228, 321–330.

Chapter 5

EFFECT OF HEAT ON GLUTEN

Costas E. Stathopoulos and Quan V. Vuong
School of Environmental and Life Sciences,
University of Newcastle, Australia

ABSTRACT

The effect of processing, and heat in particular, on the wheat gluten proteins can be difficult to explain due to its complex, and often unusual, rheological and biochemical properties. Heat denaturation of wheat gluten proteins and the accompanying rheological changes, together with a number of interactions, such as hydrogen bonding, SS bonding and hydrophobic interactions have an effect on the native structure of the protein. During the heat treatment of gluten, denaturation, aggregation and cross-linking all combine to give rise to a series of changes that affect rheological and biochemical properties alike. Different components of gluten might exhibit different responses to heat treatment, based on their parent wheat variety, their size, their composition, or the environment the heat treatment took place. Today's food scientists are yet to fully understand all the interactions and mechanisms involved in the effect of heat on gluten but this field of research has grown enormously over the last few decades and continuously expands offering us a better insight and understanding.

INTRODUCTION

Gluten proteins have been isolated for more than 250 years and have been studied extensively over the last century. It is the great importance that world populations place on the nutritional contribution of wheat, together with the prominent position wheat has as an ingredient in today's ever-expanding food industry that make understanding the effect of processing on wheat a priority for current research. Heat denaturation of wheat gluten proteins and the accompanying rheological changes are of considerable importance in the bread making process (Schofield et al., 1983). Protein denaturation has been defined as any modification in conformation (secondary, tertiary, or quaternary), not accompanied by the rupture of peptide bonds in the primary structure (Cheftel et al., 1985). A number of interactions, such as hydrogen bonding, SS bonding and hydrophobic interactions maintain the native structure of the protein. Changes in conformation (or unfolding) of protein molecules generally increases the exposure of reactive groups as well as hydrophobic regions of the protein, which are buried within their structures (Boye et al., 1997).

In concentrated protein systems, such as gluten, formation of SS bonds is the major factor leading to protein insolubility when heated (Schofield et al., 1983; Nakai and Li-Chan, 1989). The effect of heat may be manifested in two ways which appear consecutively: by unfolding of the molecule thus leading to exposure of hydrophobic sites, and by aggregation which, particularly through hydrophobic interactions and SS bonding leads to a reduction in solubility. As Clark (1992) summarised, there are three distinct events that could be identified during heat treatment, denaturation, aggregation and cross-linking. For gluten in particular the final result is the irreversible formation of a set macromolecular network.

EFFECT OF HEAT ON GLUTEN RHEOLOGICAL PROPERTIES

The heating of gluten causes changes in its physicochemical and functional properties. Most rheological studies on dough and gluten have found similar changes occurring in both during heating. LeGrys et al. (1981) and Dreese et al. (1988) have observed that the effect of heating on whole gluten mirrors that occurring in the parent dough, with a steady decrease in G' and G" up to 55°C followed by a steep increase beyond 55°C. Heat denaturation of wheat gluten results in a loss of baking quality, with the major changes occurring at 55-75°C (Schofield et al., 1983; Schofield et al., 1984). Other researchers however have suggested that heating at much higher temperatures (up to 90°C) is necessary in order to achieve a significant change in rheological properties of a gluten – water dough (Dreese et al., 1988). Davies (1986) carried out temperature sweeps for glutens containing different amounts of starch, and concluded that those samples with the lowest starch content exhibited a significant increase in their elastic modulus at about 80°C, suggesting the formation of a protein network. Hayta and Schofield (2005) observed that heating gluten above 50°C, strengthened the structure as evidenced by the increase in the elastic modulus, and they also noted that glutenin seemed to be more affected than gliadin.

According to Schofield et al. (1983) glutenin was indeed affected more by heat treatment. The reason for that is an increase in the number of rheologically effective cross-links at temperatures above 50°C, due to SH/SS interchange reactions, which are facilitated through temperature-dependent unfolding of the three dimensional structures of the proteins. The effect of these reactions is to increase the molecular size of the glutenin aggregates and perhaps to "lock" the unfolded proteins in the denatured state. As more sophisticated techniques become available, more subtle differences could be found. Changes in the dynamic rheological spectrum were also observed by Tsiami et al. (1997a, b), but at lower temperatures of 40-50°C. They claimed that the changes were due to the increase of the hydrophobic linkages of gluten or to an early denaturation of the protein. Guerrieri et al. (1996) have also observed heat induced changes in gluten's surface hydrophobicity from as early as 45°C, although no evidence of these changes could be seen by electrophoresis. Extractability in various solvents is also greatly affected by heat as demonstrated by various researchers (Schofield et al., 1983; Weegels et al., 1994a, b; Guerrieri et al., 1996; Hayta and

Schofield, 2005). Such decreases in extractability are due mainly to the decrease in solubility of the glutenin fraction for temperatures up to 90°C. α–, β–, and γ– gliadins become unextractable only above 90°C, while ω–gliadins, which are deficient in sulfur amino acids, remain extractable after heating at these temperatures (Schofield et al., 1983).

This effect was also confirmed by the findings of Attenburrow et al. (1994) who observed at approximately 90°C a very large increase in G′, presumed to be due to extensive cross-linking of the proteins. Changes observed by Lefebvre et al. (1994) when heating gluten beyond 55°C, could indicate increasing permanent cross-linking between glutenin polymers.

Stathopoulos et al. (2006) demonstrated that the elastic character of gluten protein fractions of varying MW increased with heating, with the low MW fractions exhibiting the greatest increases. Various workers have found that the low MW glutenins and gliadins have different thermal stability than the high MW glutenin polymers (Schofield et al. 1983; Weegels et al. 1994a, b; Kieffer and Wieser 2004; Kovacs et al. 2004). Because gluten is a highly polydisperse polymer system comprising a range of MW from LMW gliadins around 10^4 to HMW glutenins beyond 10^8 (Stevenson and Preston 1996; Carceller and Aussenac 2001; Lemelin et al. 2005), heating during baking is likely to have very different effects on such different polymer sizes and structures. Since the molecular weight distribution of gluten is thought to be one of the main determinants of dough rheological properties (Southan and MacRitchie 1999), and different varieties are known to have differences in MW distributions (Lemelin et al., 2005; Stathopoulos et al., 2006), the effect of heat on MW distribution within gluten from different varieties has also been examined (Stathopoulos et al. 2008). Dobraszczyk et al. (2003) showed bubble stability and extensional strain hardening for doughs from a number of wheat varieties decreasing with temperature, with the poorer varieties giving an onset of bubble instability at lower temperatures (40-50°C) than the than the stronger varieties (>60°C). Stathopoulos et al. (2008) also observed different behaviour between glutens extracted from flours with differing bread-making qualities, with gluten from the weakest flour having the highest tan δ values over the whole temperature range, indicating a greater degree of viscous behaviour. According to this research glutens from all wheat varieties examined all had a rheological transition, shown by a rapid decrease in tan δ, at about 60°C. This transition was attributed to gelatinisation of residual starch which was not completely removed during the washing of the dough and has been shown to be >10%, (Khatkar et al. 1995). The tan δ vs temperature plots for the bread-making varieties examined, were different from those of the biscuit making varieties, in that there was an additional transition in the plots for those glutens at 35 – 40°C. Such a transition has also been observed by Tsiami et al. (1997). Wang et al. (2001) also observed changes in gluten structure as early as 40°C, and suggested that hydrogen bonds were broken by the heating process, in effect the secondary structure of the proteins opening up and releasing more accessible polar groups. Such an interpretation could explain the decrease of the tan δ values of the gluten samples as a result of the increase in inter-molecular interactions.

Despite some findings pointing towards a correlation between baking volume and rheological behaviour of gluten, therefore presenting researchers with the potential to use rheology as predictive tool for baking quality, no statistically significant correlation was found between baking volume and rheological parameters measured by Stathopoulos et al. (2008).

When high temperatures are combined with high pressure treatment the effects on the rheological properties of gluten are quite dramatic. According to Kieffer et al. (2007) there is an initial increase of the gluten strength observed, however, with temperatures approaching baking conditions (80°C) the application of high pressure eventuated to complete loss of cohesivity. These results were in agreement with earlier observations by Apichartsrangkoon et al. (1999).

GLIADINS

The S-rich gliadin components are known to be heat stable due to their compact conformations (Tatham and Shewry, 1985). Gliadins also act as plastisisers reducing the stiffness of gluten (Graveland and Henderson, 1987)

Heating gliadins in the range of 50-75°C, has been found to decrease G' slightly, followed by a sharp increase in G' above 75°C. That increase was attributed to a cross-linking / aggregation reaction occurring among the gliadin molecules resulting in the formation of a network structure (Schofield et al., 1984). Kokini et al., 1995, carried out frequency sweeps at 70°C and demonstrated that gliadin began to cross-link through the formation of SS bonds. Temperature sweeps combined with frequency sweeps at given temperatures split the thermal transition onto three stages: entangled polymer flow up to 70°C, a reactive state, which results in crosslinking up to 130°C, and finally a softening of the cross-linked material (Kokini et al. 1995).

Falcão-Rodrigues et al. (2005) have suggested that since gluten development is caused by the breakage and reformulation of sulphur bridges, the higher the energy needed to perform denaturation the more difficult will be the interaction between gliadins and glutenins. They observed that interactions of the protein fractions of the wheat variety showing the lower bread-making became more difficult and consequently the gluten appeared as a weaker one. Multivariate analysis of their results suggests that it is the gliadin fraction that is mostly responsible for the lower bread making ability of the weaker wheat variety.

GLUTENINS

A number of researchers have pointed out that glutenin is more susceptible to heat treatment than gliadin (Schofield et al, 1983, Tsiami et al., 1997 a, b, Hayta and Schofield, 2005). Frequency and temperature sweep experiments with glutenin have shown that a reaction zone begins at about 90°C, in which a network starts to form due to SS bonding, reaching a maximum in structure build up due to heat setting at about 135°C. As the G' values were at a maximum at this temperature while the G" values were at a minimum, the reaction was complete (Kokini et al., 1995).

Tsiami et al., (1997) have employed a salt fractionation technique which resulted in fractions with increasing molecular weights. The glutenin subfractions exhibited an increase in G' as frequency increased, indicating protein association. The first transition was observed at about 40-50 °C, which may be a result of hydrogen bonding and hydrophobic interactions. Singh and MacRitchie (2004) found that for aqueous dispersions of gluten heated at temperatures below 100°C only HMW glutenins appear to polymerise, whilst at higher temperatures gliadins are polymerised directly into glutenins on heating without any intermediate changes. Lagrain et al. (2008) have suggested a heat induced gliadin-glutenin polymerisation explained through a SH-SS exchange mechanism. Interestingly, Hargreaves et al. (1995) have demonstrated that heat effects on the molecular flexibility of gluten can be reversible; however, this is not the case when dealing purified glutenin subunits, the difference was attributed to their differences in cross-link organisation.

EFFECT OF HEAT ON THE BIOCHEMICAL PROPERTIES OF GLUTEN

Rheological and biochemical properties of gluten are very much interconnected when it comes to heating, as the structure of gluten is mainly stabilised by SS bonds, hydrogen bonds and hydrophobic interactions (Weegels et al., 1994a, b; Weegels and Hamer, 1998; Hayta and Schofield, 2005). Schofield (1986) showed that SH/SS interchange could be facilitated by the heat-induced unfolding of the proteins. That unfolding also allows non-polar residues to be exposed making the contact of the molecule with water more difficult hence reducing significantly the solubility in aqueous media.

Protein solubility can be affected by amino acid composition and sequence, molecular weight, conformation and the contents of polar and non polar groups in amino acids. Any alteration of protein solubility as a result of heating is evidence of conformational changes in the structure of protein (Phillips et al., 1994). It is well known that exposing proteins to high temperatures for only a short period of time generally causes them to undergo conformational changes, the most visible sign of which is a decrease in extractability (Schofield and Booth, 1983; Hoseney, 1986). Thermal motion contributes to new associations and sequentially converts polypeptides into aggregates and finally into unextractable components (Autran et al., 1989). A decrease in the extractability of heat treated gluten in 60% ethanol was observed by Jeanjean et al. (1980), probably due to the formation of new SS bonds, while Pence et al., (1953) found acetic acid extracted proteins to be more sensitive to heat denaturation. Kieffer et al. (2007) have evaluated extractability of gluten, as well as various gliadin and glutenin groups, in a number of solvent at a range of pressures and concluded that cysteine containing proteins were affected when extracted with ethanol.This effect was attributed to the cleavage of intrachain disulphide bonds and their rearrangement into interchain bonds.

Extractability in other buffers, such as the strong anionic detergent SDS also decreased with the heat treatment of the protein at temperatures above 55°C (Schofield et al., 1983). Singh (2005) noted a decrease in extractable protein with high temperature and time of baking, while Stathopoulos et al. (2008) also observed decreased gluten protein extractability in SDS, with more pronounced changes at higher temperatures. That decrease in extractability suggests that heat-induced aggregation of gluten proteins occurs. The extractability in SDS of the different protein classes of gluten indicated a relation with the observed changes in functional properties, as the extracted protein from heated gluten contained a decreased level of LMW subunits and an even more extensively decreased amount of HMW subunits. At the same time, for gliadins, albumins and globulins no significant changes in extractability were observed (Schofield et al., 1983, Weegels et al., 1994a). Nagao et al., (1981) also observed a decrease in the solubility of protein from oxidised doughs after heating. Examination by gel filtration chromatography of the SDS-extracted protein from heated gluten indicated once more that glutenin was more susceptible to heat than gliadin and so became less extractable (Schofield et al., 1983; Booth et al., 1980; Hayta and Schofield, 2005). Extractability in SDS was affected when heating was carried out in the presence of various redox agents, in the work of Lagrain et al. (2006). DTT resulted in extractability lower than the control gluten protein suspension. In contrast, both potasisium iodate and potassium bromate increased extractability suggesting greater heat resistance of the gluten proteins in their presence, and less gliadin-glutenin cross-linking.

SH GROUPS AND SS BONDS

Heating gluten has been reported to result in biochemical changes that are reflected by a decrease in extractability. Along with observations on the SH groups, these observations suggest that heat setting is accompanied by changes in the SS structure of the gluten (Schofield et al., 1983, He and Hoseney 1991, Weegels et al., 1994).

Booth et al., (1979) explained heat effects on gluten between 55-75°C by a polymerisation mechanism occurring through an increase and scrambling of SS bonds. Schofield et al., (1983), and Weegels et al., (1994a), have suggested that SH/SS interchange reactions are facilitated by temperature-dependent unfolding of the three dimensional structure of protein. This in turn leads to the formation of aggregates due to the crosslinking of the unfolded proteins, hence increasing the molecular weight of the proteins.

This suggestion is supported by Lefebvre et al., (1994), who showed that glutenin polymer sizes increased significantly for samples heated above 50°C. Bloksma (1972) showed a reorganisation of intramolecular covalent bonds between subunits, while Anno (1981) concluded that crude glutenin was more sensitive to heat than gliadin and that heat-induced aggregation of gluten was due to formation of interchain SS links by an SH/SS interchange mechanism.

Schofield et al. (1983) found that total free SH groups remained constant irrespective of temperature, and observed a shift of free SH groups from an SDS-extractable form to an SDS-unextractable form. The free SH groups were predominantly in LMW glutenin aggregates and gliadin. These researchers have proposed that gluten proteins unfold up to 75°C, facilitating SH/SS interchanges between exposed groups. They explained the mechanism of gluten polymerisation on the basis of SS scrambling rather that oxidation of SH groups to SS bonds. Hansen et al. (1975) and Davies et al. (1991) have also drawn similar conclusions, suggesting that the heat setting of proteins arises through an increase in branching through SS bond formation. The authors suggested the formation of new protein cross-links between cysteine residues exposed by temperature-induced unfolding, and they proposed that it is the number and accessibility of cysteine residues that determine differences in tensile properties among glutens from different wheat varieties.

Stathopoulos et al. (2008) suggested that regardless of the parent wheat variety all glutens examined contained the highest amounts of SH groups when unheated, while the lowest amounts were observed at the highest temperature; interestingly gluten extracted from the weakest flour had much lower values throughout the temperature spectrum when compared with both glutens extracted from a commercial mixture and from breadmaking varieties, especially at higher temperatures. These observations would indicate no matter how slight the heating, it always has an effect on the number of disulphide (SS) bonds formed. That is in agreement with the effect of heat on rheology described earlier, as for most glutens there was a rheological change occurring even at 40°C.

HYDROPHOBICITY

Another approach in assessing the effect that heating has on gluten, is to measure its hydrophobicity with the use of the fluorescent probe ANS. Surface hydrophobicity (ANS binding) should decrease if the proteins unfolded as a result of the heating and then aggregated through hydrophobic interaction, thus reducing the number of ANS binding sites. The surface hydrophobicity would be expected to decrease because it is enthalpically favourable (Guerrieri et al. 1996).

Gluten proteins contain several amino acids with hydrophobic side chains (tryptophan, alanine, leucine, phenylalanine, isoleucine, valine, and proline). Also, the hydrophobic chains of longer polar side chains (lysine, glutamic acid) may also contribute to the hydrophobic interactions, which increase with heating, up to certain temperatures (Lasztity, 1984). A technique involving the fluorescent probe, 8-anilino-1-naphthalene sulfonate (ANS), has been used with gluten and it was found that heat lead to a decrease in hydrophobicity at a moisture content of 13%. Again this was attributed to protein aggregation especially in the large glutenin polymers. On increasing the moisture level, the aggregation state became constant (Weegels et al., 1994b).

Guerrieri et al. (1996) examined the binding of ANS by acetic acid extractable glutens, and suggested the presence of low and high affinity hydrophobic binding sites. Similar experiments have been carried out in other studies (Hayta and Schofield, 2005; Stathopoulos et al., 2001) but no basis for division of ANS binding sites into only two sites of high and low affinity (hydrophobicity) was found. Hayta and Schofield (2005) and Stathopoulos et al. (2001) have also found, however, that acetic acid-extractable proteins from good and poor bread making flours differ in hydrophobicity. The molecular configuration, the intermolecular bonds and the positioning of hydrophobic residues affect significantly the properties of the proteins (Zayas, 1997).

Stathopoulos et al. (2008) used the method of Guerrieri et al. (1996) to examine six glutens from varying bread-making quality flours and their results indicated that all had quite similar surface hydrophobicity values in their native (unheated) states. Most glutens showed a gradual decrease in surface hydrophobicity with further heating, and for all the lowest values were observed when they had been heated at 90°C. Such decreases could be an indication of unfolding and denaturation of the proteins. For glutens extracted from a bread making variety an increase in surface hydrophobicity on heating to 40°C was observed. That would imply an increase in the number of sites available for ANS binding within the molecule, hence an increase in the unfolding of the protein that led to more hydrophobic sites being revealed. Further heating of the gluten led to progressive decreases in surface hydrophobicity. On the other hand, heat changes to the gluten extracted from the weakest wheat variety examined seemed to have been concluded by heating at 60°C, while for a commercial 'extra-strong' flour mixture changes were not complete by reaching 90°C (Stathopoulos et al., 2008).

POLYPEPTIDE COMPOSITION AND CONFORMATION

In the work of Schofield et al. (1983) it was shown by starch gel electrophoresis and SDS-PAGE of propan-1-ol extracted proteins, that the only proteins appear unaffected by heat treatment were ω-gliadins, which do not contain cysteine residues.

Later, Menchovska et al. (1987) heated gliadins to temperatures of 100°C and 200°C in order to mimic the temperature of the crumb and the crust in bread making. Using acid-PAGE and RP-HPLC, it was shown that the more hydrophobic fractions (α,β, and γ gliadins) undergo major changes and seem to be more susceptible to heat treatment than the less hydrophobic gliadins (comprising mainly ω-gliadins). Tatham et al. (1987) investigated the effect that heat has on the conformation of low molecular weight glutenin subunits. It was found that their circular dichroism (CD) spectra were more similar to those of α,β, and γ-gliadins than those of either ω-gliadins or HMW glutenin subunits. However, no differences in the thermal stabilities of the low molecular weight fraction were found that could explain the bread making quality differences between cultivars. It is possible that this is due to resistance of gliadins to conformational changes when heated (at a moisture level of above 20%), whereas glutenin subunits suffered extensive conformational modification (Weegels et al., 1994b).

As a consequence of those conformational changes and extractability levels the distribution of the molecular weights of gluten proteins would be increasing with heating. It can be expected that an increase in MW of the polymer would be observed with the heat treatment. The hypothesis tested by Stathopoulos et al. (2008) was that as the gluten proteins became less extractable in SDS with increased heat-treatment, the high MW polymers will not be present in the extracts, giving proportionally more monomeric gliadin, which would be indicated in their MW distribution profile. The protein polymers that were not extractable would be present as the ultra high molecular weight polymers.

Indeed, according to Stathopoulos et al. (2008) the determination of the MW distribution of the six extracted gluten showed an apparent decrease of the high MW polymers as a percent of injected protein in all varieties. It was observed also that there was an increase of the unextractable gluten. In this case, there was a shift of polymer from the soluble stage to insoluble, confirming a polymer transition. The heat induced changes detected with the Flow-FFF in tandem with MALLS and SDS solubility showed that there is a heat induced transition at 40 to 50°C for the gluten polymer. Changes at this temperature range have been reported previously in the literature (Schofield et al. 1984; Guerrieri et al. 1996; Tsiami et al 1997; Dobraszczyk et al., 2003). The changes in solubility were gradual and the heat induced changes continue in steps up to 70°C. At 90°C there was a sudden change of the gluten structure (aggregation or change in conformation) where 90% of the polymer became insoluble, leaving mainly the low MW fraction, gliadin in solution. The changes in solubility and the shift of the extractable polymer to higher or lower than expected MW was also observed by Tsiami et al. (2000), where the effect of glutathione was monitored.

CONCLUSION

Overall, the effect of heat on gluten is very important for the bread making process. A number of interactions take place, the most important being the formation of SS bonds. It is these bonds that lead to protein insolubility when heated. The effect of heat on the gluten proteins may be manifested in two ways, by unfolding (leading to exposure of previously hidden hydrophobic sites) and / or by aggregation (through hydrophobic interactions and SS bonding that leads to a reduction in solubility). Glutenins are more susceptible to heating than gliadins, while extractability in various solvents is also greatly affected. That decrease in extractability, even at relatively low temperatures is an indication of conformational changes in the gluten polymer.

Previous work has shown the effects of heating on gluten: decreased solubility, increase in elastic character, increased MW, changes in SH groups, surface hydrophobicity and extractability and heat induced aggregation; some of those effects can be variety dependant. This raises the question whether these heat induced changes occur gradually with heating, or is there a sudden transition at a certain temperature, as indicated by rheological studies. Furthermore, do these changes differ between varieties of varying baking performance, and how do these changes correspond to changes in molecular weight distribution?

Can we eventually use the knowledge we have accumulated about the effect of heat on gluten to predict technological characteristic such as baking quality?

REFERENCES

Anno, T. Studies on heat induced aggregation of wheat gluten. *J. Jap. Soc. Food Nut.* 1981 34, 127-132.

Apichartsrangkoon, A.; Bell, A.E.; Ledward, D.A.; Schofield, J.D. Dynamic viscoelastic behavior of high-pressure-treated wheat gluten. *Cereal Chem.*, 1999, 76, 777-782.

Attenburrow, G.; Barnes, D.J.; Davies, A.P.;Ingham, S.J. Rheological properties of wheat gluten. *J. Cereal Sci.*, 1990, 12, 11-14.

Autran, J.C.; Galterio, G. Associations between electrophoretic composition of quality characteristics, and agronomic attributes of durum wheat. II. Protein–quality associations. *J.Cereal Sci.*, 1989, 9, 195-215.

Bloksma, A.H., Flour composition, dough rheology and baking quality. *Cereal Science Today*, 1972 17, 380-386.

Booth, M.R.; Bottomley, R.C.; Ellis; J.R.S.; Malloch, G.; Schofield, J.D.; Timms, M.F. The effect of heat on gluten physicochemical properties. *Ann.Technol.Agric.* 1980, 29, 399-408

Booth, M.R.; Melvin, M.A. Factors responsible for the poor bread making quality of high yielding European wheat. *J. Sci. Food Agric.*, 1979, 30, 1057- 1064.

Boye, J.L.; Ma, C.Y.; Harwalkar, V.R Thermal denaturation and coagulation of proteins. In *Food proteins and their applications* (S. Damodaran and A. Paraf, eds), Marcel Decker, 1997, NY, 25-44.

Carceller, J.L.; Aussenac, T. Size characterisation of glutenin polymers by HPSEC-MALLS. *J.Cereal Sci.* 2001, 33,131-142

Cheftel, J.C.; Cuq, J.L.; L'Orient, D. Amino acids, peptides and proteins. In: *Food Chemistry* (Ed. O.R.Fennema), Marcel Decker, 1985, NY, pp. 45-68.

Clark, A.H. Gels and gelling. In: *Physical chemistry of foods* (H.G. Schwartzberg and R.W. Hartel, eds) IFT Basic Symp. series, Marcel Decker, 1992, NY, pp.263-305.

Davies, A.P. Protein functionality in bakery products. In: *Chemistry and Physics of Baking* (Eds J.M.V. Blanshard, P.J.Frazier and T.Galliard). The Royal Society of Chemistry, 1986, London, pp. 89-104.

Davies, A.P.; Ingman, S.J.; Attenburrow, G. Plasticisation and mechanical properties of heat set wheat gluten. In: *Gluten Proteins 1990*. (Eds W. Bushuk and R.Tkachuk). AACC, 1991, St. Paul, MN, pp. 21-28.

Dreese, P.C.; Faubion, J.M.; Hoseney, R.C. Dynamic rheological properties of flour, gluten, and gluten-starch doughs. I. Temperature dependent changes during heating. *Cereal Chem.*, 1988, 65(4), 348-353.

Dobraszczyk, B.J.; Smewing, J.; Albertini, M.; Maesmans, G.; Schofield, J.D.. Extensional rheology and stability of gas cell walls in bread doughs at elevated temperatures in relation to breadmaking performance. *Cereal Chemistry* 2003, 80, 218-224.

Falcão-Rodrigues, M.M.; Moldão-Martins, M.; Beirão-da-Costas, M.L. Thermal Properties of glutens of two soft wheat varieties. *Food Chem.* 2005, 93, 459-465.

Graveland, A.; Henderson, M.H. Structure and functionality of gluten proteins. In: *Proc. Int. Workshop Gluten Proteins* (Eds. R. Laszitity and F. Bekes), World Scientific, 1987, Singapore, pp. 238-246.

Guerrieri, N.; Alberti, E.; Lavelli, V.; Cerletti, P. Use of spectroscopic and fluorescence techniques to assess heat induced modifications of gluten. *Cereal Chem.* 1996, 73(3), 368-374.

Hansen, L. P.; Johnston, P.H.; Ferrel, R. E. Heat moisture effect on wheat flour. I. Physical-chemical changes of flour proteins resulting from thermal processing. *Cereal Chem.* 1975, 52, 459-472.

Hargreaves, J.; Popineau, Y.; Le Meste, M.; Hemminga, M.A. Molecular flexibility of wheat gluten proteins subjected to heating. *FEBS Letters*, 1995, 372, 103-107.

Hayta, M.; Schofield, J. D., Dynamic rheological behavior of wheat glutens during heating. *J SciFood Agri*, 2005, 85: 1992–1998

He, H.; Hoseney, R.C. Gas retention in bread dough during baking. *Cereal Chem.* 1991, 68, 521-525.

Hoseney, R.C. Component interaction during heating and storage of baked products. In: *Chemistry and Physics of Baking*. (Eds. J.M.V.Blanshard, P.J. Frazier, and T. Galliard). The Royal Society of Chemistry, 1986, London, pp. 216-228.

Jeanjean, M.F.; Damidaux, R.; Feillet, P. Effect of heat ttreatment on protein solubility and viscoelastic properties of wheat gluten. *Cereal Chem.*, 1980, 57, 325-331.

Khatkar, B.S.; Bell, A.E.; Schofield, J.D. The dynamic rheological properties of glutens and gluten sub-fractions from wheats of good and poor bread making quality. *J. Cereal Sci.* 1995, 22, 29-44.

Kieffer, R.;Schurer, F.; Köhler, P.; Wieser, H. Effect of hydrostatic pressure and temperature on the chamical and functional properties of wheat gluten: Studies on gluten, gliadin and glutenin. *J Cereal Sci.*, 2007, 45, 285-292.

Kieffer, R.; Wieser, H. Effect of high pressure and temperature on the functional and chemical properties of gluten. In: Lafiandra, D, Masci, S., D'Ovidio, R. (Eds.), The Gluten Proteins. Royal Society of Chemistry, 2004, Cambridge, pp. 235-238.

Kokini, J.L.; Cocero, A.M.; Madeka, H. State diagrams help predict rheology of cereal proteins. *Food Technol.* 1995, 49, 74-81.

Kovacs, M.I.P.; Fu, B, X.; Woods S.M.; Khan K. Thermal stability of wheat gluten protein: its effect on dough properties and noodle texture. *J. Cereal Sci.* 2004, 39, 9-19.

Lagrain, B.; Thewisse, B.G.; Brijs, K.; Delcour, J.A. Mechanism of gliadin-glutenin cross-linking during hydrothermal treatment. *Food Chem.* 2008, 107, 753-760.

Lagrain, B.; Brijs, K.; Delcour, J.A. Impact of redox agents on the physico-chemistry of wheat gluten proteins during hydrothermal treatment. *J Cereal Sci.* 2006, 44, 49-53.

Lasztity, R. The Chemistry of Cereal Proteins. CRC press, 1984, Boca Raton, FL, pp.121-131

Lefebvre,J.; Popineau,Y.; Cornec, M. Viscoelastic properties of gluten proteins: Influence of prolamin composition and of temperature. In: *Gluten Proteins 1993*, Assoc. Cereal Research, 1994, Detmold, pp. 180-189.

LeGrys, G.A.; Booth, M.R.; Al-Baghdadi, S.M. The physical properties of wheat proteins. In: Pomeranz, Y., Munck, L. (Eds.), Cereals: A Renewable Resource, Theory and Practice. AACC, 1981, St. Paul, Minnesota, pp. 243-264.

Lemelin, E.; Aussenac, T.; Violleau, F.; Salvo, L.; Lein, V. Impact of variety and environment in size characteristics of wheat proteins using asymmetrical flow field-flow fractionation and multi-angle laser light scattering. *Cereal Chem.* 2005, 82, 28-33.

Menchovska, M.; Lookhart, G.L.; Pomeranz, Y. Changes in the gliadin fraction during breadmaking: isolation and characterization by high-performance liquid chromatography and polyacrylamide gel electrophoresis. *Cereal Chem.* 1987; 64, 311-314.

Nagao, S.; Endo, S.; Tanaka, K. Scanning electron microscopy studies of wheat protein fractions from doughs mixed with oxidants at high temperatures. *J. Sci. Food Agric.*, 1981, 32, 235-242.

Nakai, S.; Li-Chan, E. Effects of heating on protein functionality. In: *Protein quality and the effects of processing* (R.D. Phillips and J.W.Finley, eds) Marcel Decker, 1989, NY, pp. 125-144.

Pence, J.W.; Mohammad, A.; Mecham, D.K. Heat denaturation of wheat gluten. *Cereal Chem.*, 1953, 30, 115-126.

Phillips, L. G.; Whitehead, D.M.; Kinsella, J.E. Protein stability. In: *Structure Function Properties of Food Proteins*. Academic Press, 1994, NY, pp. 25-46.

Schofield, J.D.; Bottomley, R.C.; Timms, M.F.; Booth, M.R. The effect of heat on wheat gluten and the involvement of sulphydryl-disulphide interchange reactions. *J. Cereal Sci.* 1983, 1, 241-253.

Schofield, J.D.; Bottomley, R.C.; LeGrys, G.A.; Timms, M.F.; Booth, M.R. Effects of heat on wheat gluten. In: *Gluten Proteins, Proceedings of 2nd International Workshop on Gluten Proteins* (Eds. A. Graveland, and J.H.E. Moonen) TNO, 1984, Wageningen, pp. 81-90.

Schofield J.D.; Booth, M.R. Wheat proteins and their technological significance. In: *Developments in Food Proteins* (Ed. B.J.F. Hudson) Applied Science Publishers Ltd., 1983, Barking, Essex, pp. 1-65.

Schofield, J.D. Flour proteins:structure and functionality in baked products. In: Chemistry and physics of baking (J.M.W. Blanchard, P.J.Frazier and T. Galliard, eds) The Royal Society of Chemistry, 1986, London, pp 14-29.

Singh, H. A study of changes in wheat protein during bread baking using SE-HPLC. *Food Chem.* 2005, 90, 247-250.

Singh, H.; MacRitchie, F. Changes in proteins induced by heating gluten dispersions at high temperature. *J.Cereal Sci.* 2004, 39, 297-310.

Southan, M.; MacRitchie, F. Molecular weight distribution of wheat proteins. *Cereal Chem.* 1999, 76, 827-836.

Stathopoulos, C.E.; Tsiami, A.A.; Schofield, J.D. Rheological and size characterisation of gluten proteins from cultivars varying in baking quality. *Aspects of Applied Biology 64,* 2001 *Wheat Quality,* 181-183.

Stathopoulos, C.E.; Tsiami, A.A.; Dobrasczcyk, B.J.; Schofield, J.D. Effect of heat on rheology of gluten fractions from flours with different bread-making quality, *J.Cereal Sci.* 2006, 43, 322-330.

Stathopoulos, C.E.; Tsiami, A.A.; Schofield, J.D.; Dobrasczcyk, B.J. Effect of Heat on Rheology, Surface Hydrophobicity and Molecular Weight Distribution of Glutens Extracted From Flours with Different Bread-Making Quality *J. Cereal Sci.* 2008, 47, 134-143.

Stevenson, S.G.; Preston, K.R. Flow field flow fractionation of wheat proteins. *J. Cereal Sci.* 1996, 21, 121-131.

Tatham, A.S.; Field, J.M.; Smith, S.J.; Shewry, P.R. The conformations of wheat gluten proteins. II. Aggregated gliadins and low molecular weight subunits of glutenin. *J. Cereal Sci.* 1987, 5, 203-214.

Tatham A.S.; Shewry, P.R. The conformation of wheat gluten proteins. The secondary structures and thermal stabilities of α, β, γ and ω- gliadins. *J. Cereal Sci.* 1985, 3, 103-112.

Tsiami, A.A.; Bot, A.; Agterof, W.G.M.; Groot, R.D. Rheological properties of glutenin subfractions in relation to their molecular weight. *J. Cereal Sci.* 1997a, 26, 15-27.

Tsiami, A.A.; Bot, A.; Agterof, W.G.M. Rheology of mixtures of glutenin subfractions. *J. Cereal Sci.* 1997b, 26, 279-287.

Tsiami, A.A.; Every D.; Schofield J.D. Redox reactions in dough: effects on molecular weight of glutenin polymers as determined by Flow FFF and MALLS. In: Wheat Gluten eds. PH Shewry and AS. Tatham, The Royal Society of Chemistry, 2000, Cambridge, pp 244-248.

Wang, Y.; Belton, P.S.; Bridon, H.; Garanger, E.; Wellner, N.; Parker, M.L.; Grant, A.; Feillet, P.; Noel, T. Physicochemical properties of carubin: A Gluten-like protein *J. Agric. Food Chem.,* 2001, 49, 3414-3419.

Weegels, P.L.; Hamer, R.J. Temperature induced changes of wheat products. In: *'Interactions: the keys to cereal quality'* (Eds. R.J.Hamer and R.C. Hoseney) AACC, 1998, St Paul, MN, USA.

Weegels, P.L.; Veroek, J.A.; de Groot A.G.M.; Hamer, R.J. Effect on gluten of heating at different moisture contents. I. Changes in functional properties. *J. Cereal Sci.* 1994a, 19, 31-38.

Weegels, P.L.; Veroek, J.A.; de Groot, A.G.M.; Hamer, R.J. Effect on gluten of heating at different moisture contents. I. Changes in physicochemical properties and secondary structures. *J. Cereal Sci.* 1994b, 19, 39-47.

Zayas, J.P. Functionality of proteins in Food. Springer-Verlag, NY, 1997.

In: Gluten: Properties, Modifications and Dietary Intolerance
Editor: Diane S. Fellstone
ISBN: 978-1-61209-317-8
©2011 Nova Science Publishers, Inc.

Chapter 6

MODERN CONCEPTS PATHOGENESIS OF CELIAC DISEASE: FROM GLUTEN TO AUTOIMMUNITY

Asma Ouakaa-Kchaou[*]
Department of Gastroenterology and Hepatology,
Habib Thameur Hospital, Tunis-Tunisia

ABSTRACT

Background: Celiac disease, also known as gluten-sensitive enteropathy and non-tropical sprue, is a prevalent autoimmune disorder that is triggered by the ingestion of gluten and related prolamins in genetically susceptible individuals. The classic celiac lesion occurs in the proximal small intestine with histological changes of intestinal villous atrophy, crypt hyperplasia, intraepithelial lymphocytosis and leukocyte infiltration of the lamina propria. The pathogenic mechanisms in this disease are not yet well understood, but it is clear that genetic, environmental and immunological factors play a role.
Aim: To provide an evidence-based overview of the pathogenesis of celiac disease.
Methods: Review based on relevant medical literature.
Results: Celiac disease is uniquely characterized by a defined trigger (gluten proteins from wheat and related cereals), the necessary presence of HLA-DQ2 or HLA-DQ8, and the generation of circulating autoantibodies. Celiac disease has become one of the best-understood HLA-linked disorders. Well-identified haplotypes in the human leukocyte antigen (HLA) class II region (either DQ2 [DQA*0501-DQB*0201] or DQ8 [DQA*0301-DQB1*0302]) confer a large part of the genetic susceptibility to celiac disease. The immune response in celiac disease involves the adaptive, as well as the innate, and is characterized by the presence of anti-gliadin and anti-transglutaminase antibodies, lymphocytic infiltration in the epithelial membrane and the lamina propria, and expression of multiple cytokines and other signaling proteins.
Conclusion: Gluten-free diet is currently the only effective mode of treatment for celiac disease; nevertheless, there is a growing demand for alternative treatment options.

---

[*] Corresponding author: Ouakaa-Kchaou Asma, Department of Gastroenterology. Habib Thameur Hospital, Address: 8, Ali Ben Ayed Street Montfleury 1008, Tunis - Tunisia., Phone: (00216) 98 383 053, Fax: (00216) 71 493 167, E-mail: asma.kchaou@àliov.fr or asma.kchaou@rns.tn

Better understanding of the mechanism of the disease is likely to add other choices for therapy in the future.

INTRODUCTION

Celiac disease (CD; also called gluten sensitive enteropathy or celiac sprue) is a frequently occurring disorder that is precipitated in genetically susceptible individuals by the ingestion of proline-rich and glutamine-rich proteins in wheat, rye, and barley. The development of agriculture, which started in the Middle East about 10,000 years ago, not only led to the development of ancient civilizations but also resulted in radical changes in the composition of the human diet. One of those changes was the introduction of cereal based food products and today such food products are very common in a normal diet. Researchers postulate that the condition first developed after the last ice age in the fertile crescent of the Middle East with the cultivation of grains, and the first description dates from the 1st and 2nd centuries [1]. CD is a chronic inflammatory disease characterized by flattened villi on the small bowel mucosa. It has a diverse clinical heterogeneity, and increases both morbidity and mortality. Knowledge of many aspects of this disorder, in particular pathogenesis is inadequate. As far as pathogenesis is concerned, CD represents a unique and privileged model since both an external trigger, the gluten peptides, and the autoantigen, the ubiquitous enzyme tissue transglutaminase (tTG), have been identified. CD, has several autoimmune features, of which the highly disease-specific autoantibodies to the enzyme tTG are particularly striking. The production of these antibodies in CD is turned on and off depending on exposure to dietary gluten. However, despite the great advances of the last decade in understanding the molecular mechanisms of mucosal lesions, our knowledge about the immune recognition of gluten and the consequent immune response is still far from being complete. Better understanding of the mechanism of the disease is likely to add other choices for therapy in the future.

This review investigates the scientific literatures about modern pathogenesis applied for celiac disease. In order to do that, a search of articles related to celiac disease using MEDLINE/PubMed, Text Books and other scientific publications was accomplished. English-language articles, published until April 2010 were selected.

CAUSATION

The balance of evidence suggests that the celiac immunopathology involves a complex individualized interplay of many pathophysiological variables on a genetic background [2]. Celiac disease develops as a consequence of the encounter between an environmental trigger (derivatives of gluten from wheat, rye, and barley), immunologic factors, and a genetically predisposed host, with the possible participation of other environmental cofactors. However, less is known about the early mechanisms and initiating steps that lead to celiac disease. How and when gluten sensitivity and development of autoimmunity first occur is unknown. It is also still debated which of two principal pathogenic mechanisms could cause the flat lesion, negative effects on the surface epithelium that cause cell damage and loss of enterocytes (villous atrophy) followed by compensatory crypt hyperplasia or positive effects on the crypt cells directly inducing proliferation, with villous atrophy being only 'apparent' as a result of crypt hyperplasia. There are considerable experimental data to support both these possibilities, and the lesion most likely reflects a 'joint venture'.

Triggers

Over 50 years ago, the crucial role of wheat gluten in the development of celiac disease was discovered. For the vast majority of human beings, cereals represent an important source of nutrients, whereas for CD patients certain cereal products represent poisons that not only destroy small intestinal mucosa, but also predispose to gastrointestinal malignancy. Gluten is the water-insoluble material from wheat flour [3]. It is a rubbery mass that consists of storage proteins that remain after starch is washed from wheat-flour dough. These proteins have different solubilities in alcohol-water solutions and, thus, can be roughly separated into two fractions gliadins and glutenins. Gluten proteins have a complex chemistry and are responsible for the baking properties of wheat-water absorption capacity, cohesivity, viscosity, and dough elasticity [4]. The major components of gluten are glutenins and gliadins, both representing families of proteins. Gluten molecules are storage proteins and they have high concentrations of glutamine and proline residues (35 and 15% of the total amino acid content). While the glutamine provides nitrogen to the developing seedling, the proline may be involved in the protection of the seeds against drought. Glutenins contain multiple cysteine residues that allow the formation of covalent bonds between glutenin molecules. These bonds establish the matrix of dough and its elasticity. Moreover, gliadins can link to this matrix and enable water binding, which is important for the viscosity of the dough [3]. Storage proteins, with a similar aminoacid composition to the gliadin fractions of wheat, have been identified in barley and rye, and show a close relation to the taxonomy and toxic properties of wheat cereal that affect people with celiac disease. When collectively considered, the alcohol soluble fractions of cereals are designated as prolamins, a term reflecting the particular amino acid composition that is a high content of proline and glutamine and, depending on the cereal, they have been termed secalin for rye, hordein for barley, avenin for oats other than gliadin for wheat [5]. Oats are thought to activate CD only rarely, and, consistent with this, oat avenins are more distantly related to the analogous proteins in wheat, rye, and barley and have a substantially lower proline content [6]. The gluten peptides and the related prolamins are responsible for triggering mucosal lesions in CD patients. Among them, only gliadin has been investigated in great detail. Analysis of gliadin has identified more than a hundred components that have been grouped into four main types (ω5-, ω1,2-, α/β-, γ-gliadins), on the basis of their electrophoretic mobility at acidic pH or, more modernly, into three major types according to their N-terminal amino acid sequence, designated as α-, γ- and ω- types [5]. Even if the correspondence between the old and the new nomenclature is not complete, it may be assumed that the electrophoretically separated α- and β-gliadins mainly constitute the α-type gliadins, while γ- and ω-gliadins are comprised into the γ- and ω-types, respectively. Glutenins can be divided into groups of high molecular weight and low molecular weight [3,4]. Immunogenicity and toxicity in the high-weight group have been shown. Gluten and other proline-rich proteins are poorly digested in the normal human small intestinal tract due to a lack of prolyl endopeptidases. This results in the generation of gluten peptides that can be as large as 10–50 amino acids in length. Gluten is also rich in the amino acid glutamine [7]. Current understanding indicates that different gluten peptides are involved in the disease process in a different manner, some fragments being 'toxic' and others 'immunogenic'. Specifically, a fragment is defined 'toxic' if it is able to induce mucosal damage either when added in culture to duodenal mucosal biopsy, or when administered *in vivo* on proximal and distal intestine, whereas a fragment is defined 'immunogenic' if it is able to specifically stimulate HLA-DQ2- or DQ8-restricted T cell lines and T cell clones derived from jejunal mucosa or peripheral blood of celiac patients [5]. Although several gluten epitopes are immunostimulatory, some are more active than others. An immunodominant peptide of 33 aminoacids (residues 57−89) identified from an α-gliadin fraction has functional properties attributable to many proline and glutamine residues [7]. Proline gives the peptide increased resistance to gastrointestinal proteolysis (in people with and without celiac disease) [4]. In addition, the repetitive presence of these residues makes the peptides a preferred substrate of tTG, whose main function is to catalyze the covalent and irreversible cross linking of a glutamine residue in glutamine-donor proteins with a lysine residue in glutamine-acceptor proteins which results in the formation of an ε-(γ-glutamyl)-lysine (isopeptidyl) bond [5]. However, apart from cross linking its substrates, tTG can also hydrolyze peptide-bound glutamine to glutamic acid either at a lower pH or when no acceptor proteins are available, a process leading to an enhanced immunogenicity of gluten peptides [4,5]. Furthermore, these peptides naturally adopt a structural configuration characterized by a left-handed polyproline II helical conformation, which is the one preferred by all bound HLA class II ligands, strengthening the binding with HLA-DQ2 and HLA-DQ8 molecules on antigen-presenting cells.

Genetic Susceptibility

The pathogenesis of CD is firmly rooted in host genetic factors. Celiac disease is one of the most common genetically based diseases. This was first evident from clinical observations of multiple cases of CD within families (the risk of celiac disease in first-degree relatives being 20–30 times greater than in the general population), and the high (approximately 70-85%) rate of concordance for CD among monozygotic twins. It is known that CD is associated with specific MHC class II alleles that map to the HLA-DQ locus. Moreover, the presence of specific HLA-DQ alleles is necessary, although not sufficient, for the phenotypic expression of CD in virtually all affected individuals, irrespective of geographic location. Indeed, almost all individuals with biopsy-confirmed CD express HLA-DQ alleles that encode specific HLA-DQ2 heterodimers or specific HLA-DQ8 heterodimers [4,5,9,10]. Most patients (90-95%) carry a variant of DQ2 (alleles DQA1*05/DQB1*02) and others carry a variant of DQ8 (alleles DQA1*03/DQB1*0302). Rare DQ2-/DQ8- patients carry alleles that code for one chain of the DQ2-encoded heterodimer (DQA1*05 or DQB1*02) [11,12]. Because more than 98% of people with celiac disease share the major histocompatibility complex II class HLA-DQ2 or HLA-DQ8 haplotype, the inclusion of HLA typing for these haplotypes is useful, especially in patients with equivocal biopsy results or negative serological tests, or for patients already on a gluten-free diet. People who do not have HLADQ2 or HLA-DQ8 haplotypes are unlikely to have celiac disease [9]. The association between HLA genes (COELIAC1 locus on chromosome 6p21) and celiac disease is very strong compared with other HLA-linked diseases; however, researchers estimate that the genetic effect attributable to HLA is 40 to 50%. Moreover, DQ2 is carried by roughly a third of the general population, thus suggesting that HLA is only partly the cause of the condition. The concordance rate between HLA-identical siblings (sib recurrence risk for celiac disease is 10%) is much lower than between monozygotic twins; thus, other non-HLA regions must be involved [9]. Candidate genetic regions that possibly increase CD susceptibility have been noted in some populations on chromosomes 2, 3, 4, 5, 6 (telomeric of the HLA locus), 9, 11, 18, and 19 [6]. However, the putative risk posed by these regions is substantially lower than that posed by expression of the HLA-DQ2 or HLA-DQ8 CD susceptibility alleles. Furthermore, the genes in these regions and the possible mechanisms by which they might contribute to disease susceptibility currently are not known, although some of these regions also have been associated with other autoimmune or inflammatory diseases. This additional susceptibility might be conferred by: COELIAC2 (5q31−33), which contains cytokine gene clusters; COELIAC3 (2q33) that encodes the negative costimulatory molecule CTLA4; and COELIAC4 (19p13.1), which contains the myosin IXB gene variant encoding an unconventional myosin that alters epithelial actin remodeling [3,4,6,8,11].

Environmental Cofactors

Some drugs can have a role in enhancing a person's susceptibility to gluten. A course of interferon α could activate celiac disease in predisposed people [4]. Intestinal infections might cause a transient rise in small-bowel permeability and could lead to up-regulation and release of tissue transglutaminase that, in turn, enhances gluten immunogenicity. Rod-shaped bacteria have been identified in the intestinal epithelium in children with celiac disease, although this colonization could just be coincidental [4]. Rotavirus infections could raise the risk of celiac disease in genetically predisposed children. The homology between the rotavirus-neutralizing protein VP-7 and tissue transglutaminase might explain how rotavirus infection is implicated in the development of celiac disease [4]. It has also been hypothesized that, at least in some individuals, recent surgery may result in compromised epithelial barrier function and the initiation of intestinal inflammation. Timing of the introduction of gluten in infancy was demonstrated to be a considerable factor. Changes in infant-feeding practices might affect the rise and fall of the disease. The introduction of dietary gluten while infants were still being breastfed, and the introduction of small or medium amounts rather than large amounts, are protective factors against the disease in early and perhaps later childhood. Early introduction of cereals into the infant diet before 3 months may be associated with an increased risk of developing childhood celiac disease, but there is no evidence that delaying the introduction of gluten into the diet of children at high risk for celiac disease beyond the 3- to 6-month period is beneficial. However, large follow-up studies are, therefore, needed to clarify how dietary cofactors affect the development of this condition before a child's immunity is established and to identify primary prevention strategies [4].

Autoimmune Features of Celiac Disease

Autoantibodies

Untreated CD patients usually have increased levels of antibodies against agent wheat gluten, several other food antigens, and autoantigens present in the mucosa. The autoantibodies in CD patients are primarily directed against the $Ca2^{+}$ activated form of transglutaminase (TG2), although antibodies to calreticulin and actin are also found [10]. The antibodies to TG2 are both IgA and IgG isotypes, but the IgA antibodies, which are primarily produced in the intestinal mucosa and which show signs of somatic mutation, demonstrate the highest disease specificity [4,9]. It is uncertain whether autoantibodies play a role in the pathogenesis of celiac disease. A weak inhibitory effect of the antibodies on certain TG2 catalyzed reactions has been reported [9]. This might be relevant to lesion formation. TG2 is involved in the formation of active TGF-β by cross linking of the TGF-β binding protein. Indirect inhibition of TGF-β activation can have broad effects, including dysregulation of enterocytes and immune cells. In addition to its enzymatic activity, TG2 is involved in the attachment and motility of fibroblasts and monocytes via interactions with integrins and fibronectin [13]. Hence, CD villous atrophy could be caused by TG2 autoantibodies disturbing the migration of fibroblasts and epithelial cells from the crypts to the tips of the villi [10]. It is possible that autoantibodies to TG2 are involved in the extraintestinal manifestations of CD. A particularly interesting observation is the finding that some patients with severe liver disease have undetected CD, and hepatic failure is reversed when the patients are put on a gluten-free diet [13]. Another interesting observation comes from CD patients with dermatitis herpetiformis. It appears that these patients, in addition to their anti-TG2 antibodies, have antibodies that target TG3, a transglutaminase that is uniquely expressed in the dermal papillae of dermatitis herpetiformis patients [10]. The mechanism underlying the formation of the autoantibodies in CD is not understood. The production of the anti-TG2 IgA antibodies is likely to be dependent on cognate T cell help to facilitate isotype switching of autoreactive B cells. Autoreactive T cells specific for TG2 could provide the necessary help for B cell production of anti-TG2 IgA but, as TG2 is expressed in the thymus; such cells are not likely to survive thymic selection. An alternative explanation could be that the complexes of gluten and TG2, in the form of thiolester-linked enzyme substrate intermediates, permit the gluten-reactive T cells to provide help to the TG2-specific B cells by a mechanism of intramolecular help. This model can explain why the serum TG2 antibodies in CD disappear when the patients are put on a gluten-free diet. When the gluten goes, so does the T cell help needed for the B cells to switch isotype and differentiate to plasma cells [13].

Autoreactive Intraepithelial Lymphocytes

Normal intraepithelial lymphocytes (IELs) are tightly regulated by inhibitory and activating NKG2 natural killer (NK) receptors that recognize non-classical HLA molecules induced by IFN-γ (HLA-E) and stress (MIC) on intestinal epithelial cells [13]. Untreated CD is characterized by an increased density of proliferating $TCR\alpha\beta^{+}$ $CD8^{+}$ $CD4^{-}$ and $TCR\gamma\delta^{+}$

CD8⁻ CD4⁻ cells in the villous epithelium, which upregulate the selectively activating NKG2 receptors [11]. In contrast to the number of TCRαβ⁺ CD8⁺ IELs, which returns to normal when gluten is removed from the diet, the number of TCRγδ⁺ IELs remains at an elevated level [11]. Interestingly, the Vg1Vd1 gdT cell subset, which is increased in CD patients, recognizes MIC molecules expressed by stressed enterocytes [13]. Upregulation of NKG2 receptors might favor the elimination of stressed enterocytes by IELs either by reducing the TCR activation threshold or by mediating direct killing. NKG2 receptors serve as co-stimulatory or coactivating molecules that can render celiac IELs reactive to low avidity antigens. Importantly, NKG2D receptors expressed on CD IELs can also mediate direct cytotoxicity of intestinal epithelial cells expressing MIC in an antigen non-specific manner [13]. The upregulation of these NKG2 receptors seems to be driven by IL-15 [13], which is expressed by CD enterocytes. IL-15 has a role in NKG2D-mediated autoimmunity. IL-15 seems to play a critical role in the expansion of IELs, and in the induction of MIC molecules on intestinal epithelial cells [13]. The mechanism leading to increased IL-15 expression by intestinal epithelial cells in CD is poorly understood; however, there is evidence that, in CD gut biopsies and in the absence of CD4⁺ T cell activation, the α-gliadin p31–43 peptide induces MAP kinase activation in enterocytes and upregulates IL-15 expression [13], suggesting that it might trigger a yet undefined signalling pathway.

PATHOPHYSIOLOGY

Study of the pathogenesis of celiac disease has focused on the mechanisms by which gluten peptides, after crossing the epithelium into the lamina propria, are deamidated by tissue transglutaminase and then presented by DQ2⁺ or DQ8⁺ antigen-presenting cells to pathogenic CD4⁺ T cells. Once activated, the CD4⁺ T cells drive a T-helper-cell type 1 response that leads to the development of celiac lesions, namely intraepithelial and lamina propria infiltration of inflammatory cells, crypt hyperplasia, and villous atrophy [4].

Epithelial Translocation of Gluten Peptides and Enterocyte Processing of Gliadin Epitopes

Under normal physiologic conditions, the intestinal epithelium, with its intact intercellular tight junctions, serves as the main barrier to the passage of macromolecules, such as gluten proteins [10]. The primary event of a gluten-induced immune response requires epitope-bearing oligopeptides to have access to lamina propria inside the relatively impermeable surface of the intestinal epithelial layer. Ordinarily, such oligopeptides are efficiently hydrolyzed into amino acids, di- or tri-peptides by peptidases located in the brush border membrane of the differentiated enterocytes before they can be transported across the epithelium [5]. In fact, in the intestinal lumen, gastrointestinal proteases are a first defense against potentially toxic dietary proteins, including the incompletely digested gluten proteins. However, it remains unknown how very small amounts of gluten peptides can enter the intestinal mucosa of genetically susceptible subjects to initiate the inflammatory cascade. Gut permeability is enhanced in celiac disease and gluten can reach the lamina propria through different routes. Investigators have postulated that there is a paracellular route on the basis of amplified expression of zonulin, a protein implicated in the opening of tight junctions and T-helper-1-induced changes in the expression, localization, or phosphorylation of epithelial junctional proteins in active disease [4,5]. However, the paracellular passage of gluten is not proven, whereas study results have shown that the immunodominant α2-gliadin-33mer16 translocates into the lamina propria via an interferon-γ-dependent transcytosis, which suggests involvement of the transcellular route. Furthermore, the protected transport pathway, which is driven by retrotranscytosis of secretory IgA through transferrin receptor CD71, promotes the influx of intact, and thus harmful, gluten peptides [4,5].

Modification and Presentation of Gluten Peptides: The Role of Tissue Transglutaminase

HLA-DQ2 and HLA-DQ8 heterodimers can bind and subsequently present "gluten" peptides to populations of $CD4^+$ T cells in the lamina propria of the small intestine. How HLA-DQ2 and HLA-DQ8 bind such peptides was an enigma for several years, because the peptide-binding groove of HLA-DQ2 and HLA-DQ8 favors the binding of peptides with negatively charged residues at key anchor positions. Such negatively charged amino acids are largely absent from native "gluten" peptides generated in the human intestinal tract. However, this puzzle was solved after the discovery that the target antigen of an autoantibody present in many patients with CD was a tissue transglutaminase [14]. It is a calcium-dependent, ubiquitous enzyme that catalyses post-translational modification of proteins and is released during inflammation. TG could have at least two crucial roles in celiac disease: as the main target autoantigen for antiendomysial antibodies and t-TG antibodies, and as a deamidating enzyme that raises the immunostimulatory effect of gluten [4,15]. Expression and activity of tissue transglutaminase are raised in the mucosa of patients with celiac disease, where, by deamidating glutamine to glutamic acid, this enzyme makes gliadin peptides negatively charged and therefore more capable of fitting into pockets of the, DQ2/DQ8 antigen-binding groove. Additional functions of the enzyme in celiac disease consist of cross-linking gluten peptides, thus forming supramolecular complexes, and catalyzing either the binding of gluten peptides to interstitial collagen 44 or the incorporation of histamine into gluten proteins (transamidation). All of these actions contribute to the formation of a wide range of T-cell-stimulatory epitopes that might be implicated in different stages of the disease. The α2-gliadin-33mer fragment is the most immunogenic because it harbors six partly overlapping DQ2-restricted epitopes.

Effector Mechanisms

The immune response to gliadin takes place in two compartments, the lamina propria and the epithelium. While the part played by lamina propria CD4 T cells in pathogenesis is recognized, the role of CD8 T cells in the intestinal epithelium (intraepithelial lymphocytes) is controversial.

GLUTEN AND THE ADAPTIVE IMMUNE RESPONSE

In active celiac disease, the lamina propria is expanded in volume, which is due, in part, to recruitment of T lymphocytes, plasma cells and dendritic macrophages expressing HLA molecules, ICAM-1 and CD25 (IL-2 receptor α-chain), an infiltrate indicative of a T cell-mediated immune response [14,15]. The role of lamina propria CD4 T cells is lent support by the strong genetic linkage to HLA-DQ2 and HLA-DQ8 and the identification of DQ2 and DQ8 anti-gluten restricted T cells that secrete interferon γ [9,10]. The presence of an increased amount of gluten peptides in the lamina propria together with a genetic predisposition greatly increases the risk of breaking the oral tolerance toward these peptides. In the mucosa of patients with active celiac disease, gluten-reactive $CD4^+$ T cells produce several proinflammatory cytokines, interferon γ being dominant that trigger various effector mechanisms including raised secretion of tissue-damaging matrix metalloproteinases and heightened cytotoxicity of intraepithelial lymphocytes against enterocytes with increased enterocyte apoptosis and villous flattening [16,17]. Several aspects of the molecular mechanisms that drive the immune response in celiac disease are unknown. Some proinflammatory cytokines are upregulated in active celiac mucosa, namely interferon γ, interferon α, interleukin (IL) 6, interleukin 18, and interleukin 21; however, paradoxically, tumor necrosis factor α, the most powerful promoter of inflammation, and interleukin 12, the main cytokine that primes T cells for interferon γ production, are not raised. T-cell-mediated immunity alone does not account for the expansion of cytotoxic $CD8^+$ intraepithelial lymphocytes.

GLUTEN AND THE INNATE IMMUNE RESPONSE

Intraepithelial lymphocytosis is a hallmark of celiac disease and its importance is shown by the major complications of the disease, refractory sprue and enteropathy-associated T-cell lymphoma, which represent expansions of abnormal intraepithelial lymphocytes. The innate immune response is the initial activation step induced by α-gliadin peptides and is mediated primarily by $CD8^+$ T cells, enterocytes, macrophages, and dendritic cells [10]. The majority of intra-epithelial lymphocytes are CD8 positive and express natural killer markers such as CD94, suggesting that they may be cytotoxic to enterocytes. A smaller percentage of these lymphocytes are both CD4/CD8 negative and express the primitive γ/δT cell receptor. Unlike the CD8 intra-epithelial lymphocytes or lamina propria infiltrate, this population does not regress on gluten withdrawal. It has been proposed that these γ/δ lymphocytes form part of innate rather than acquired immunity. They do not appear to require HLA for antigen recognition and recognize stress proteins such as MHC class I chain-related gene (MIC) A and MIC-B expressed on epithelial cells, subsequently recruiting polymorphs and monocytes [16]. It is well known that under inflammatory and/or stress stimuli, enterocytes express not only HLA class II molecules, but also MIC and HLA-E molecules that are recognized by the natural killer receptors NKG2D and CD94 present on intraepithelial lymphocytes (IELs) [5]. The recent discoveries that IL-15, a key regulatory cytokine which supports the homeostasis between innate and adaptive immunity, may be produced by intestinal epithelial cells and is a potent stimulant of IELs, and that gliadin causes an overexpression of MIC-A in the intestinal mucosa of CD patients have focused the attention on IL-15 as a key molecule in the pathogenesis of CD. Interleukin 15 is implicated in activating perforin-granzyme-dependent cytotoxicity by celiac intraepithelial lymphocytes, and in promoting their expression of the natural-killer receptors CD94 and NKG2D, thus contributing to enhanced enterocyte apoptosis. Furthermore, interleukin 15 might have a crucial role in the emergence of T-cell clonal proliferations because of its antiapoptotic action on the intraepithelial lymphocytes, therefore predisposing patients to the malignant complications of celiac disease [4]. Some gluten peptides can directly induce mucosal damage via a non-T-cell-dependent pathway (innate response). The best characterized peptide is the nonimmunodominant p31-43/49 fragment of α-gliadin that is thought to be unable to stimulate gluten-reactive $CD4^+$ T cells. In celiac mucosa, the toxic gliadin peptide 31–43 is able to induce an innate immune response by up-regulating the expression of IL-15, cyclo-oxygenase-2, and the activation markers CD25 and CD83 in lamina propria cells having the characteristics of macrophages, monocytes, and dendritic cells, without binding to HLA-DQ2 or -DQ8 and without stimulating $CD4^+$ T cells. Interleukin 15 production, in turn inhibits the immune-regulatory signaling of transforming growth factor β, promotes dendritic cell maturation, and causes epithelial stress [4,5]. Autoantibodies against tissue transglutaminase might contribute to mucosal damage by preventing the generation of the active form of transforming growth factor β, or through inducing enterocyte cytoskeleton changes with actin redistribution via their interaction with the extracellular-membrane-bound tissue transglutaminase, or by stabilizing tissue transglutaminase in a catalytically advantageous conformation. Conversely, because the enzyme autoantibodies might inhibit the activity of tissue transglutaminase, they could block their pathogenic role. However, inhibition is far from complete, and the residual enzyme activity could be sufficient to exert its pathogenic role.

FUTURE DIRECTIONS: AN ALTERNATIVE TO THE GLUTEN-FREE DIET?

In celiac patients there is a permanent intolerance to gluten. The treatment is a life-long gluten-free diet to prevent chronic enteropathy and reduce the risk of developing gastrointestinal lymphoma or carcinoma. However, the dietary restriction is inconvenient to patients and some gluten-free products are relatively unpalatable, which may lead to poor compliance or inadvertent intake of gluten. In addition, some patients do not respond rapidly or completely to dietary restriction and require treatment with immunosuppressive agents [16,17]. Some aspects of celiac disease pathogenesis are unknown, including the relation between the events in the epithelium, the contribution of innate or adaptive immunity, the role of regulatory T cells, and the possible function of gliadin peptides as ligands for mammalian pattern-recognition receptors. Nevertheless, improved understanding of the molecular basis of celiac disease has enabled researchers to suggest alternatives to a gluten-free diet. These novel treatments are aimed at blunting the immune stimulatory effects of gluten. However, we emphasize that some of these drugs (tissue-transglutaminase-inhibitors and monoclonal antibodies) have a poor safety profile, and their hypothetical use could, thus, be reserved for complicated forms of the disease. Alternative approaches to manage the condition aim to eliminate detrimental gluten peptides through genetic detoxification or enzyme modification.

Further characterization of epitopes within gluten may provide sufficient information to breed out these toxic sequences from wheat, whilst retaining its baking properties. Given the complexity of wheat genetics and the observation that toxic epitopes are scattered throughout the gluten genome, this is unlikely to be achieved by conventional breeding techniques. There are also concerns about the safety and ethics of genetic modification, but this is an option that is currently being explored [16]. The development of grains that have low or no immunotoxicity can be achieved with transgenic technology that deletes or silences harmful gluten sequences or selectively breeding non-toxic wheat varieties. However, although the availability of non-toxic bread might resolve problems such as poor palatability and the high cost of gluten-free foods, patient's social problems would nonetheless remain [4]. Gluten can be detoxified within the intestine by oral administration of prolyl endopeptidases, enzymes that cleave the proline-rich and glutamine-rich immunostimulatory gluten peptides into small sequences with reduced toxic effects. These approaches are limited by the need for complete digestion of the toxic epitopes of gliadin, which has never been shown. These strategies should be investigated in large long-term clinical studies to assess safety and clinical effectiveness. The development of gluten-sensitive rhesus macaques could be a promising model of celiac disease for testing these new treatments [4]. The mechanisms for induction of oral tolerance in humans remain poorly understood. Restoring immunological tolerance to gluten would represent the ideal cure for celiac disease. There has been much interest in the concept of oral tolerance in immune mediated disease, whereby an oral antigen is administered in an attempt to desensitize auto-reactivity against self-antigens. In animal models, such strategies have been shown to prevent the development of some autoimmune diseases, but none have been able to modulate established disease. The timing and nature of antigenic exposure, together with the immunological status of the host, appear to be critical in influencing whether sensitization or tolerance develops. Continued research in celiac disease may increase our understanding of the mechanisms that result in loss of tolerance and allow us to identify those individuals with the potential to develop the condition. This knowledge could then be used to design a regimen of gluten epitope exposure that did not result in celiac disease. In established disease, successful immunotherapy is likely to prove more difficult. However, if gluten-specific T cells could be inactivated or deleted, tolerance to gluten should be restored. In principle, such immunomodulation could provide a selective, tailored treatment of celiac disease, although this is likely to be many years in the future [16].

CONCLUSION

Celiac disease is a T-cell mediated chronic inflammatory bowel disorder with an autoimmune component. Loss of tolerance to gluten is a possible cause. Although such loss of tolerance has been identified in people without the disease, as manifested by the presence of peripheral blood T cells that respond to gluten, the development of an anti-gluten T-cell response in the intestine is specific to people with celiac disease. The reason for this occurrence is obscure; however, changes in intestinal permeability secondary to alterations in intercellular tight junctions or in the processing of gluten are potential mechanisms. As a common autoimmune disorder, celiac disease represents a unique model of autoimmunity because of the identification of a close genetic association with HLA-DQ2 and HLA-DQ8, a highly specific humoral autoimmune response (autoantibodies against the autoantigen, tTG), and, most important, the external trigger, gluten peptides. Because of the genetic predisposition to celiac disease, an individual's intolerance to gluten is lifelong and self-perpetuating. The amount of gluten ingested, the gene dose of HLADQ2 and HLA-DQ8 (homozygous individuals appear to be at highest risk of celiac disease), and the local expression of tTG appear to be important determinants of celiac disease manifestation and severity.

REFERENCES

[1] Losowsky MS. A history of coeliac disease. *Dig Dis* 2008; 26: 112-20.
[2] Brandtzaeg P. The changing immunological paradigm in coeliac disease. *Immunol Letters* 2006; 105: 127-39.
[3] Koning F, Gilissen L, Wijmenga C. Gluten: a two-edged sword. *Immunopathogenesis of celiac disease Springer Semin Immun* 2005;27:217-32.
[4] Di Sabatino A, Corazza GR. Coeliac disease. *Lancet* 2009;373:1480-93.
[5] Ciccocioppo R, Di Sabatino A, Corazza GR. The immune recognition of gluten in coeliac disease. *Clin Exper Immunol* 2005;140:408-16.
[6] Kagnoff MF. Celiac disease: pathogenesis of a model immunogenetic disease *J Clin Invest* 2007;117:41-9.
[7] AGA Institute Medical Position Statement on the Diagnosis and Management of Celiac Disease. *Gastroenterology* 2006;131:1977-80.
[8] Ciclitira PJ, Moodie SJ. Coeliac disease. *Best Pract Res Clin Gastroenterol* 2003; 17:181-95.
[9] Green P, Jabri H. Coeliac disease. *Lancet* 2003;362:383-91.
[10] Niewinski M. Advances in Celiac Disease and Gluten-Free Diet. *J Am Diet Assoc* 2008;108:661-72.
[11] Stepniak D, Koning F. Celiac Disease, Sandwiched between Innate and Adaptive Immunity. *Human Immunology* 2006;67:460-8.
[12] Catassi C, Fornaroli F, Fasano A. Celiac disease: From basic immunology to bedside practice. *Clin App Immunol Rev2002*;3: 61-71.
[13] Sollid LM, Jabri B. Is celiac disease an autoimmune disorder? *Current Opinion Immunol* 2005;17:595-600.
[14] Reif S, Lerner A. Tissue transglutaminase, the key player in celiac disease: a review. *Autoimmun Rev* 2004;3: 40-5.
[15] Briani C, Samaroo D, Alaedini A. Celiac disease: From gluten to autoimmunity. *Autoimmunity Reviews* 2008;7:644–50.
[16] Dewar D, Pereira SP, Ciclitira PJ. The pathogenesis of coeliac disease. *Inter J Bioch Cell Biol* 2004; 36:17–24.
[17] Ciclitira PJ, Johnson MW, Dewar DH, Ellis HJ. The pathogenesis of coeliac disease. *Molecular Aspects Med* 2005; 26: 421–58.

In: Gluten: Properties, Modifications and Dietary Intolerance
Editor: Diane S. Fellstone

ISBN: 978-1-61209-317-8
©2011 Nova Science publishers, Inc.

Chapter 7

EMULSION PROPERTIES OF DIFFERENT PROTEIN FRACTIONS FROM HYDROLYZED WHEAT GLUTEN

S. R. Drago[1,3], R. J. González[1] and M. C. Añón[2,3]

[1] Instituto de Tecnología de Alimentos -FIQ- Univ. Nacional del Litoral, Santa Fe, Argentina,
[2] CIDCA UNLP-CIC-CONICET, La Plata, Argentina
[3] CONICET CONICET, CIDCA, calle 47 y 116, La Plata, Argentina

ABSTRACT

Many protein sources that are found in the market are obtained as by-products and there is a great interest in using them as protein ingredients with adequate functionality for food formulation. Structure modification allows to add value and to diversify their uses. The viscoelastic properties of wheat gluten have restricted its use in baked products, and the diversification of gluten applications depends of the improvement of its solubility in a wider pH range. An alternative for that is the enzymic hydrolysis.

The objective of this work was to evaluate emulsion properties of protein fractions obtained by extracting at 3 pH different hydrolyzed gluten samples.

Hydrolyzates were made using two commercial enzymes (acid and alkaline proteases) to reach 3 different hydrolysis degrees (DH) for each enzyme. Extracts were obtained at 3 different pHs (4, 6.5 and 9) and were diluted to a protein concentration of 4 g/l. Each extract was used to make the corn oil: extract emulsions (25:75, W/W).

Emulsion capacity was determined by measuring droplet size distribution and the stability using a vertical scan macroscopic analyzer.

Regarding emulsion capacity, multifactor ANOVA (factors: pH and DH) made for droplet size distribution parameters showed that there were no differences between samples.

Regarding stability evaluation, alkaline protease extract emulsions were more stable in particle migration phenomena by creaming, but showed higher coalescence rates than those corresponding to acid protease extract emulsions.

It was also observed that for both enzymes, as DH increases, coalescence rates decrease for the 3 pHs extracts and creaming rates increase for pH 4 and 9 extracts. In the case of pH 6.5, acid protease extracts emulsions showed a clear creaming instability by flocculation, probably due to the electrical charges suppression of the peptides adsorbed at the interface, since 6.5 is a pH near the isoelectric point of gluten proteins. It is suggested that acid protease extract emulsions showed a certain degree of bridging flocculation. This caused higher creaming rates but a lower coalescence as a consequence of the bridging. We conclude that although pH of the extraction and DH did not affect emulsion capacity, emulsion stability depended on the pH, DH and the enzyme used.

INTRODUCTION

Many protein sources that are found in the market are obtained as by-products and there is a great interest in using them as protein ingredients with adequate functionality for food formulation (Vioque *et al.* 2001).

Wheat gluten represents about 72% of wheat protein. It is a by-product of the wheat starch process, which is available in large amounts and at relatively low cost. Because it is insoluble in water at near-neutral pH and is viscoelastic when hydrated, gluten is mainly used to improve the properties of flour for bread-making and for textured food (Popineau *et al.* 2002; Babiker *et al.* 1996).

Enzymic hydrolysis is one of the alternatives that allow protein modification by means of structural transformations. The produced peptides have characteristics that depend on hydrolyzed protein (substrate), enzyme and conditions of hydrolysis (pH, temperature, E/S ratio, time) which determine the degree of hydrolysis (Spellman *et al.* 2002). Other than the decrease in the average molecular weight, cleavage is usually accompanied by increase in the number of charged groups, an exposure of reactive groups and important structural rearrangements which could to expose to the aqueous phase some hydrophobic regions originally buried within the protein molecule. These factors could modify techno-functional properties such as solubility, viscosity, gelation, emulsifying and foaming properties (Phillips and Beuchat 1981; van der Ven *et al.* 2001).

Peptides may possess new nutritional (higher digestibility), functional or biological properties (techno-functional or bio-functional properties). Enzymatic hydrolyzates are thus used in food industry, parachemistry and pharmacy (Kong *et al.* 2007).

In order to act as emulsifiers, proteins should possess a range of properties: high surface hydrophobicity, limited aggregation tendency in solution, high solubility over a large range of pH, well-balanced distribution of hydrophilic and hydrophobic domains, flexibility and strong interaction at the oil/water interface (Damodaran 1994). Nevertheless, few proteins possess all the required properties (Linares *et al.* 2001a) and it is necessary a structural modification.

It has been observed that mild protein hydrolysis increase emulsifying properties but impair the stability (Linares *et al.* 2000). Ultra-filtration has been used to fractionate gluten hydrolyzates as a way to produce a peptide fraction with optimized surface properties (Popineau *et al.* 2002). Another alternative would be the fractionation by extraction at different pH (Drago *et al.* 2008a).

Emulsions are thermodynamically unstable dispersions of two immiscible liquids, most often stabilized by surfactants. The emulsion type, so-called oil-in-water (O/W) or water-in-oil (W/O) morphology essentially depends on the physicochemical formulation (Pizzino *et al.* 2009).

Commonly, energy must be supplied to produce such metastable mixtures. Surfactants reduce the surface free energy required to increase any interfacial area, by lowering the interfacial tension, and allow finely dispersed media to be created easily (Abismail et al. 1999). The reduction of interfacial tension makes easier the interface area creation, but even though the adsorption of macromolecules slightly contributes to that, it is not so important in the emulsion formation phase. In this stage, the time and the mixing energy related to the equipment are more important than the superficial activity (Stainby 1986; Kempa et al. 2005).

Once emulsions are formed, they are subject to several forms of instability. The mainly involved mechanisms are flocculation, creaming, coalescence, and phase separation (Robins 2000). The two first are generally reversible and involve particle aggregation and migration. Coalescence is irreversible and related to particle size modification. The later implies the breaking of the emulsion.

Reversible flocculation of droplets can be followed by creaming or sedimentation, according to the respective densities of dispersed and continuous phases. The migration rate of the particles of the dispersed phase is given by Stoke's law (although within certain limits). The irreversible changes (coalescence) lead to formation of larger drops, less stable emulsions and eventually to phase separation. Phase inversion can occur with temperature, or composition change (Abismail et al. 1999).

Emulsions could be stabilized by low interfacial tension, a resistant interfacial film formed by adsorbed proteins able to oppose attraction between emulsion droplets conducing to droplet coalescence, steric hindrance and electrostatic charge that contribute to the net repulsive electrostatic interaction between particles surfaces, low average droplet size and homogeneous droplet size distribution, a high continuous phase viscosity, and a high oil volumetric fraction value (Cheftel et al. 1989).

The quality of an emulsion is related to its stability. This latter depends on several parameters, among which are formulation variables (nature and amount of stabilizing agent governing interfacial tension, viscosity of the continuous phase, density difference between continuous and dispersed phases), droplet charge and sedimentation or creaming rate, and process variables (order of mixing, power output, type of contacting apparatus, flow regime), controlling average droplet size and droplet size distribution (Abismail et al. 2000).

The aim of the present work was to evaluate emulsifying properties of peptide fractions, obtained by extracting at three different pHs hydrolyzed gluten samples with different degree of hydrolysis and from two proteolytic enzymes, and to correlate these properties with their structural characteristics.

MATERIAL AND METHODS

Commercial vital gluten provided by Molinos SEMINO S.A. (Carcarañá - Santa Fe) was used. Gluten composition was the following: moisture: 5.95% (AACC 44-15A method), protein (N x 5.7): 77.20 % d.b. (Kjeldahl –AACC 46-11 method), starch: 13.15 % d.b. (Ewers polarimetric method), ether extract 0.71 % d.b. (AACC 30-25 method) and ash: 0.834 % d.b. (ICC N° 104- IRAM N° 15851 Standard technique). In order to disperse vital gluten in water and secure a uniform suspension, a moderate thermal treatment was carried out (Drago et al. 2008). The thermal-treated gluten (TTG) sample was used in all experiments as a substrate. The acid and alkaline proteases used in the experiment were provided by GENENCOR S.A. (Arroyito-Córdoba). The acid enzyme (Ac) was a fungal protease derived from *Aspergillus oryzae* (31.000 HU/g), which is a mixture of endo/exopeptidases whose characteristics are the following: effective pH: 3.5 – 9, optimum pH: 4.3 – 5, temperature range: 30 – 50 °C. The alkaline enzyme (Al) was an endoprotease derived from *Bacillus Licheniformis*, which effective pH range was 7- 10, optimum pH: 9.5, temperature range: 25-70 °C, optimum: 60°C.

PREPARATION OF HYDROLYZATES

Hydrolyzates were prepared in a 5 l batch reactor with agitation, using a thermostatized bath. Protein concentration was 8% (W/W) and 3N HCl or NaOH were added in order to reach and maintain a constant pH. After enzyme inactivation, the hydrolyzates were frozen at -20 °C and lyophilized. Reaction parameters used for acid protease were: temperature: 55°C, pH: 4.25, Enzyme/Substrate ratio (E/S): 5%. Enzyme inactivation was carried out at 70°C for 15 min. Reaction parameters used for alkaline protease were temperature: 60°C, pH: 9.5, E/S ratio: 0.095%. Enzyme inactivation was carried out at 80-85°C for 10 min.

Hydrolyzates were obtained at three different trichloroacetic acid index (TCAI) for both acid and alkaline protease hydrolyzates: 14, 22 and 32%, named Ac14, Ac22, Ac32 for acid enzymic hydrolyzates and Al14, Al22, Al32 for alkaline protease hydrolyzates.

Hydrolysis Reaction Progress

The hydrolysis progress was followed by means of the trichloroacetic acid index (TCAI), using TCA 20% and diluting the sample in a 1:1 ratio. N was measured by the Semimicro-Kjeldahl method. The TCAI, which was used as an indirect measurement of degree of hydrolysis (DH), was calculated as follows:

TCAI = [N soluble in TCA (hydrolyzate) – N soluble in TCA (blank) x 100] / total amino N

Preparation of Fractions at Different pH

In order to obtain the hydrolyzates fractions at different pH (4, 6.5 and 9), a 2% (W/W, dry basis) solution of the different hydrolyzates was prepared (Drago and González 2001). The pH was achieved by adding 0.8N HCl or 0.8N NaOH. The samples were stirred for 1 hour at room temperature, and then centrifuged during 15 min at 8000xg at room temperature. The supernatant (the extract at each pH) was frozen and protein content determined by semimicro-Kjeldahl method.

Emulsion Formation

Oil in water emulsions (25:75, W/W) were made with 5 g corn oil and 15 g 4 g/l protein (Nx5.7) extracts at different pH, from hydrolyzed gluten and TTG samples. An Ultra-Turrax

T25 homogenizator equipment and S 25 N-10 G generator of 7.5 mm rotor diameter were used. The generator was introduced until 10 mm to the end of the 25 ml vessel, during 1 min at 20.000 rpm.

Emulsion Capacity (EC)

Emulsion capacity (ie. Emulsion forming ability) was investigated by measuring particle size distribution directly after homogenizing with a Malvern Mastersizer-X (Malvern Instruments, UK) in the range of diameters 0.4 a 2000 µm. D[4,3] and its standard deviation (SD), and D[3,2] vs. droplet diameter were determined.

Five samples with flocculation type instability visually observed (Ac14, Ac22, Ac32 extracts at pH 6.5 and Al32-6.5) and other Al extracts at 6.5 (Al14-6.5 and Al22-6.5) were also evaluated using an emulsion dilution with SDS 0.1%, in order to assure completed deflocculation of oil droplets.

The droplet-size distribution was expressed as volume distribution and defined the average particle size

$$D[4,3] = \sum_{i=1}^{N} (n_i d_i^4) / \sum (n_i d_i^3)$$

The value D[v,s] =D[3,2], called Sauter average diameter, is inversely proportional to the specific surface area of droplets (Quintana et al. 2002). The standard deviation corresponding to D[4,3] is related to the polydispersity of the system.

Emulsion Stability

An optical analyzer QuickScan (Beckman-Coulter Inc. Fullerton, CA, USA) was used. This kind of apparatus was used to study the stability of various emulsions and concentrated colloidal dispersions. The destabilization by processes such as coalescence, flocculation, creaming or sedimentation could be detect much earlier than the naked eye's operator, especially in the case of opaque and concentrated systems could see these phenomena (Lemarchand et al. 2003).

QuickSacan is a liquid dispersion optical characterization instrument where the sample to be analyzed is contained in a cylindrical glass measurement cell. The reading head of the instrument is composed of a pulsed near infrared light source (λ= 850 nm) and two synchronous detectors. The transmission detector receives the light, which goes through the sample (0°) (T), while the back-scattering detector receives the light back-scattered (BS) by the sample (135°). The head of the instrument scans the entire length of the sample (about 65 mm), acquiring transmission and back scattering data every 40 µm. The T and BS profiles are obtained as a function of the sample height. An important BS signal (and a very small one in T) is observed when dispersions are opaque (droplet rich), and it is the opposite in the case of transparent to turbid dispersion (droplet depleted) (Quintana et al. 2002).

Once a sample was placed in the instrument, both light profiles were recorded every 1 min during 1 hour.

To better visualize the signal modifications as a function of time, BS profiles at $t = 0$ were subtracted from all the others and expressed as Δ_{BS}. By means of measuring Δ_{BS} it is possible to discriminate among particle migration process (sedimentation or creaming) and the variation of the particle size (flocculation and coalescence) (Pan et al. 2004).

Creaming /Clarification Kinetics

The kinetics of creaming/clarification was followed by measuring the variation of peak thickness at a base value (Pan et al. 2002). It was determined in the zone corresponding to 6.5 - 35 mm of the tube (end to top) at -5% Δ_{BS}, as a time function, every 1 min during 1 hour.

Some emulsions prepared with extracts presented creaming instability and then reached a constant value before 60 min, but for other emulsions the kinetics is cero order, having lineal tendency until 60 min. In order to compare all samples, the slope of the lineal part during the first 20 min was determined.

Emulsions made with pH 6.5 extracts from Ac hydrolyzates presented a different behavior from the other samples. Few min after the emulsion was made a clear separation interface was observed. A clear emulsion at the bottom of the tube and an opaque one at the top were observed. This interface moved to the top along the time. For this, the evaluation for these samples was made at −25% Δ_{BS} and between 19 and 40 mm. Because the Al33-6.5 extract also presented this tendency, but the profiles were different, the evaluation was made at -10% Δ_{BS} and between 6.5-40 mm,

Coalescence Kinetics

When the instability process involve only particle migration, the Δ_{BS} graphic have a positive region corresponding to the creaming phase and a negative region in the bottom of the tube caused by clarification. When there are changes in particle size (by coalescence) the Δ_{BS} profile will be negative, because of the increased size of the particles and its decreasing density.

These emulsions presented both types of instability.

The coalescence was measured by the mean value of the negative peak between 50-60 mm (top of the tube). A decreased first order function fits to the experimental data:

$$y = y0 + A \cdot e^{-x/b} \tag{1}$$

The first derivate of (1) equation respect to x (time) allows obtaining the coalescence rate,

Coalescence rate = A x (-1/b)

Statistical Analysis

Software Statgraphics Plus 3.0 was used for statistical analysis. LSD (Least significant difference) test was used to determine statistical differences among samples ($p<0.05$).

RESULTS AND DISCUSSION

1. Degree of Hydrolysis for Thermal Treated Gluten (TTG) Hydrolyzates

Table 1 shows TCAI of hydrolyzates obtained for both enzymes at the different time of hydrolysis. It is observed that the higher TCAI reached is 32.6% for Ac. This level was the maximum selected in order to avoid impairing surface properties by an excessive hydrolysis.

Table 1. Degree of Hydrolysis of hydrolyzates. Ac: acid protease. Substrate concentration: [S]= 8%, pH=4.25, T=55°C, E/S= 5%; Al: Alkaline protease: [S]= 8%, pH=9.5, T=60°C, E/S= 0.095%.

Degree of hydrolysis	Ac (%)	Al (%)
TCAI-14%	14.1	13.03
TCAI-22%	22.5	19.7
TCAI-32%	32.6	31.5

2. Evaluation of the Capacity of Forming and Stabilizing Emulsions

2.1. Emulsion Capacity (EC)

EC vas evaluated by determining the media diameter D[4,3] from volume distribution and D[3,2] (Sauter diameter), which is inversely proportional to the created specific superficial area.

Figure 1 shows the general profile observed for D[3,2] vs. droplet for emulsions made for all extracts at pH 4, 9 and 6.5 (the later with SDS 0.1%). It is possible to observe a bimodal population.

The size of the oil droplets depends largely on the emulsifying procedure, i.e., experimental conditions, equipment design. Linares et al. (2001b) observed non-gaussian distribution curves (that indicates that several populations were present), for emulsions made with valve-homogenizer, using gluten hydrolyzate dispersions at 7.5 mg/ ml and suggested that the interfacial layer was not saturated at 15 mg/ ml. However, Agboola et al. (1998), working with high hydrolyzed whey proteins, showed that increasing peptide concentration does not decrease emulsion particle size, but contributes to the stability. Thus, a bimodal distribution not always is related with the lack of saturation of interfacial layer.

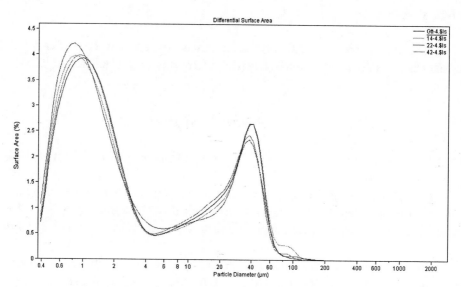

Figure 1. General profile of particle size distribution for D[3,2]: (Surface Area%) vs. particle diameter (μM).

The medium values of D[3,2] (Sauter diameter), D[4,3] and its standard deviation for Ac and Al emulsions are showed in Table 2.

The values obtained for media droplet diameter D[4,3] were similar to those observed by Popineau *et al.* (2002) for emulsions made with hydrolyzed wheat gluten and sunflower oil.

Multifactor ANOVA analysis showed that emulsions made with Ac extracts present no significant differences for D[4,3], D[3,2] and SD when the effect of degree of hydrolysis was evaluated. When pH effect was evaluated, it was observed that the samples at pH 6.5 presented significant difference for D[3,2], D[4,3] and SD. Because of a pronounced flocculation was observed, samples at pH 6.5 also were analyzed with SDS 0.1%, to avoid flock during the evaluation of droplet size distribution (Antón *et al.* 2002). When the ANOVA was made whit the results obtained with SDS, there were no significant differences for D[3,2], D[4,3] and SD, indicating that the differences initially observed were due to the flock and not to different droplets sizes.

Multifactor ANOVA analysis showed that emulsions made with Al extracts present no significant differences for D[4,3], D[3,2] and SD when the effect of degree of hydrolysis and pH were evaluated. All samples at pH 6.5 also were analyzed with SDS 0.1%, and it was possible to observed flock formation mainly for Al32, although in a less extension than for Ac-6.5 emulsions.

Figure 2 and 3 show the size droplet distribution for volume% (D[4,3]) for Ac and Al emulsions, respectively. The different profiles obtained for D[4,3] for the emulsions made with different extracts are mainly due to differences in the population of droplets with sizes higher than 70-80 μm. This is related with flocks produced by bridging flocculation and not related with the emulsion ability of different peptide extracts.

Table 2. Indexes of droplet diameter distribution

Sample	D[4,3]	D[3,2]	SD
TTG-4	36,16	11,04	20,83
TTG-9	40,5	11,29	28,57
Acid protease			
14-4	35,66	10,8	20,85
14-6.5	75,66	19,6	44,44
14-6.5-SDS	38,45	10,99	27,4
14-9	45,72	11,12	35,87
22-4	35,23	10,95	18,49
22-6.5	67,17	17,25	41,37
22-6.5-SDS	37,62	10,67	26,35
22-9	34,84	10,35	21,97
32-4	40,82	11,5	24,6
32-6.5	64,32	16,11	40,52
32-6.5-SDS	38,89	11,32	25,18
32-9	34,91	10,56	20,2
Alkaline protease			
14-4	40,27	11,83	23,18
14-6.5	36,74	9,764	24,56
14-9	32,7	9,007	21,93
22-4	36,32	10,1	21,7
22-6.5	37,37	9,637	25,25
22-6.5-SDS	29,46	8,434	15,51
22-9	32,36	8,819	21,32
32-4	34,16	10,32	18,91
32-6.5	44,4	12,61	27,73
32-6.5-SDS	35,22	10,89	21,21
32-9	32,01	10.32	20,38
32-9-SDS	32,92	10,34	17,12

In a quiescent system, adsorption of proteins at the oil-water interfaces is thought to be a diffusion-controlled process. Then, the rate of adsorption should be dependent only upon the size and shape of the molecule, viscosity of the solvent, and the temperature (Damodaran 1994). However, the higher diffusion rate of lower molecular weight is not a critical factor to having surface activity since during homogenization (turbulent condition) the collision and adsorption of the protein at interface is facilitated (Halling 1981; Phillips 1981; Damodaran 1989). Nevertheless, it was found that with a homogenizer like that used in our work, the structural characteristics of proteins (superficial hydrophobicity, flexibility, hydrophilic/ hydrophobic balance) have influence on the stabilized surface area (Kato and Nakai 1980; Voutsinas et al. 1983; Kato et al. 1983; Stainby 1986) and the amphiphilic character of the peptides could be more important than their size for emulsifying properties (Rahali et al. 2000).

Extend of adsorption and the ability to reduce interfacial tension and form a cohesive film depend on conformational stability, adaptability at phase boundaries and distribution of hydrophilic and hydrophobic groups on the protein surface. However, the effects of controlled

effects like pH and DH could be masking with others, related with the equipment for emulsifying, contributing to the great complexity of the system.

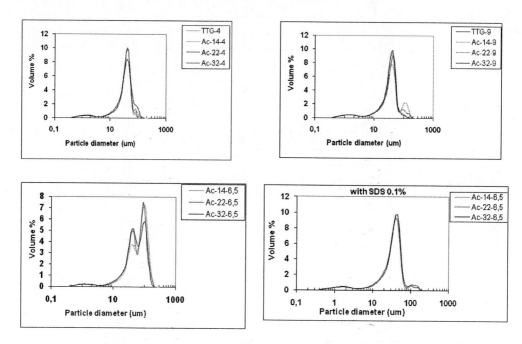

Figure 2. Size droplet distribution for volume% (D[4,3]) for emulsions made with extract at pH 4, 6.5 and 9 with thermal treated gluten (TTG) and Acid protease hydrolyzates (TCAI: 14, 22 and 32%).

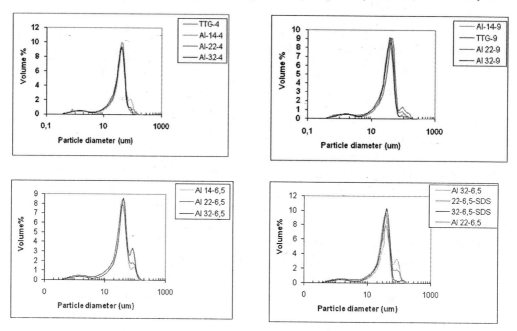

Figure 3. Size droplet distribution for volume% (D[4,3]) for emulsions made with extract at pH 4, 6.5 and 9 with thermal treated gluten (TTG) and Alkaline protease hydrolyzates (TCAI: 14, 22 and 32%).

Regarding the molecular weight of peptides in hydrolyzates, a minimum peptide length seems to be desired for good emulsion properties. However, peptide size is not the only factor influencing the emulsion behavior of peptides. Also amphiphilicity of peptides is important for interfacial and emulsifying properties of peptides (Van der Ven 2001).

2.2. Emulsion Stability

The processes that affect emulsion stability (coalescence, flocculation and creaming) drive to the separation of continuous and dispersed phases (Elizalde *et al.* 1991). These processes are not independent among them. For example, creaming favors coalescence (Stainby 1986).

Figure 4 shows Delta Back scattering (ΔBS) for TTG extract at pH 4, having the extract at pH 9 (figure not shown) the same profile than that of pH 4. It is possible to observe the instability by creaming (tube bottom) and coalescence (tube top).

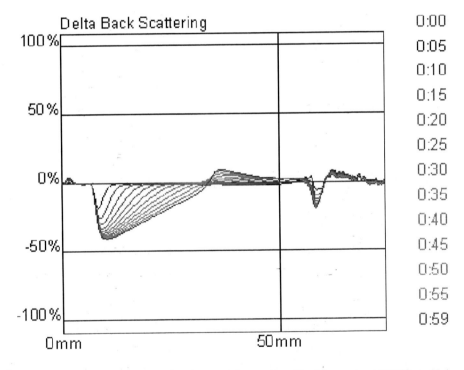

Figure 4. Difference of Back Scattering for emulsion made with gluten extract (TTG) at pH 4.

Figure 5 shows that at pH 4, extracts from Ac hydrolyzates form emulsions that are unstable by creaming and coalescence for 14% DH hydrolyzates. The profile of emulsions made with Ac22-4 is the same than that for Ac14 and it is not shown. However, for Ac32-4 coalescence it is not observed.

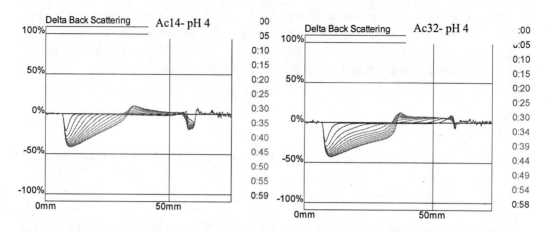

Figure 5. Difference of Back Scattering for emulsions made with Acid protease hydrolyzate extracts (Ac14 and 32) at pH 4. Ac22 showed the same profile than Ac14.

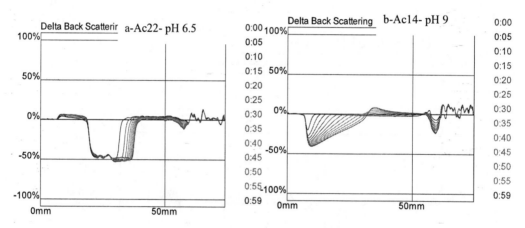

Figure 6. Difference of Back Scattering for emulsion made a- with Ac22 extract at pH 6.5. Ac14 and Ac32 showed the same profile than Ac22 at this pH; b- Difference of Back Scattering for emulsion made with Ac14 extract at pH 9. Ac22 and Ac32 emulsions showed the same profile than Ac14.

For emulsions made with Ac22 extract at pH 6.5 (Figure 6a), a profile of fast flocculation, creaming and coalescence is observed. For Ac14 and Ac32 the same profile was observed (figures not shown).

Instability profiles for emulsion made with Ac14 extract at pH 9 is shown in Figure 6.b. Creaming and coalescence were observed. The profiles of emulsions made with higher degrees of hydrolysis (Ac22 and Ac32) are the same than that obtained with Ac14.

Respect to emulsions made with Al protease extracts, at pH 4 and 9, instability by creaming and coalescence in emulsions made with the hydrolyzates at three DH was observed (Figure 7a and 7b). At pH 6.5, Al14 and Al22 show the same profile, but Al32 presented coalescence and creaming with intense flocculation (Figure 8).

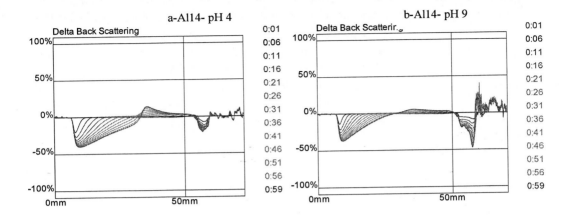

Figure 7. a- Difference of Back Scattering for emulsion made with Al14 extracts at pH 4. Al 22 and Al32 showed the same profile than Al14; b- Difference of Back Scattering for emulsion made with Al14 extract at pH 9. Al22 and Al32 showed the same profile than Al14.

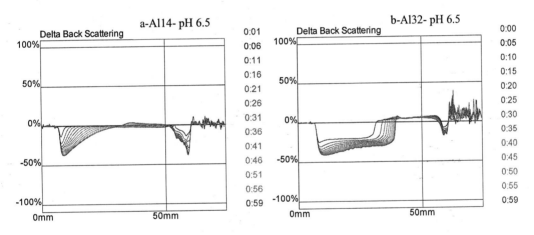

Figure 8. a- Difference of Back Scattering for emulsion made with Al14 extract at pH 6.5. Al22 showed the same profile than Al14; b- Difference of Back Scattering for emulsion made with Al32 extract at pH 6.5.

ANALYSIS OF EMULSION CREAMING RATE (CR)

Figure 9 shows the CR of emulsions made with the different extracts. Emulsions made with TTG extracts at pH 4 and 9 presented the same CR.

Emulsions made with extracts of Ac hydrolyzates had diverse behavior according the pH. At pH 4, only Ac32 have higher CR but at pH 9, a higher DH, higher CR.

Emulsions made with extracts at pH 6.5, are different from the other pHs. Once the emulsion is made, a clear interface is observed, with an opaque part at the top of the tube and a clear one in the bottom. The interface moved to the top with time. This could be due to the creaming of flocculated emulsion (Santiago et al. 2004). This means that at pH 6.5 flocculation of creaming droplets happened. Then, the flocked phase suffers coalescence (in all DH evaluated). This is coincident with the results of particle size, since it was observed an important difference when SDS was used. At higher DH, emulsions had fewer tendencies to creaming (Figure 10). The flocculation in creaming emulsion at pH 6.5 probably is due to the pH, because near to isoelectric point (pI), oil droplets flocculation is favors because of the diminution of the net charge of the polypeptides, leading to the lower repulsion and flocks formation (Tornberg et al. 1990; Dickinson 1998; Dalgleish 2006). At pH 6.5, acid protease extracts form emulsions that showed bridging flocculation. At higher DH, adsorbed peptides at the interface are lower in size, and have lower capacity to interact with other peptides adsorbed at oil droplet surface, leading to a decrease in bridging and then in flocculation capacity.

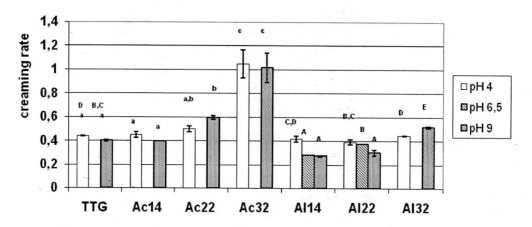

Figure 9. Creaming rate for emulsions made with different extracts, measured at first 20 min (lineal behavior). Lowercase letters: ANOVA for TTG and Ac-hydrolyzate emulsions; capital letters: ANOVA for TTG and Al-hydrolyzate emulsions. Different letters means significant differences among samples ($p<0.05$).

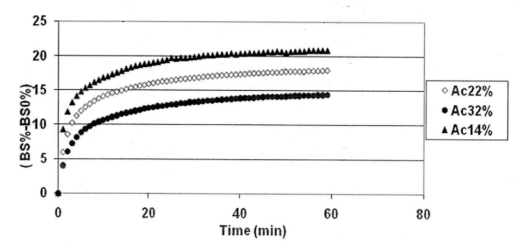

Figure 10. Difference of Back Scattering for emulsions made with Ac extracts at pH 6.5.

For the emulsions made with Ac extracts at pH 4 and 9, the emulsion remained opaque at the bottom of the tube for a time and a cream developed in the top. The cream increased in the top and then is possible to observe turbidity in the bottom (Figure 11). In this case, it is not possible to see a limit. This kind of creaming is for poly-disperse emulsions, where individual droplets or few aggregates are moving to the top (Santiago et al. 2000). This creaming is difficult to see, but it is detected using Quick Scan o Turbiscan equipment.

For emulsions made with extracts from Al hydrolyzates it was observed that at pH 4, the behavior was the same than that for Ac pH 4. The CR was higher for Al32 than the other lower DH and TTG.

At pH 6.5, at higher DH, higher the CR was, but Al32 showed flocculation like emulsions made with pH 6.5 extracts from Ac.

At pH 9, emulsions made with Al14 and Al22 hydrolyzates were more stable than TTG, being Al32 the most unstable sample.

For Al14 and Al22 emulsions made with pH 9 extracts were more stable respect to the creaming than at pH 4.

In general, emulsions made with Al extracts were more stable than those made with Ac extracts and at higher DH, the higher amount of low molecular weight components is related with higher creaming instability.

Popineau et al. (2002) observed for emulsions made with wheat gluten hydrolyzates that amphiphilicity of peptides is an important factor related with the ability of peptides for stabilizing interfaces. More hydrophobic and positively charged peptides have better emulsions properties that neutral and hydrophilic peptides.

CR depends on the difference of density between continuous and disperse phase, on molecular weight peptides, electrolytes presence and the viscosity of continuous phase (Tadros and Vicent 1983). This means that other factors than DH and pH could influence in this complex system.

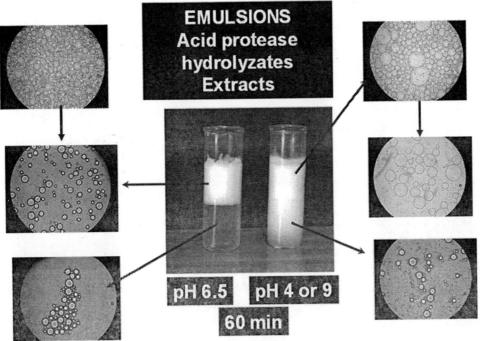

Figure 11. Emulsions made with Ac extracts after 60 min of initial preparation. Micro-photography from the different phases.

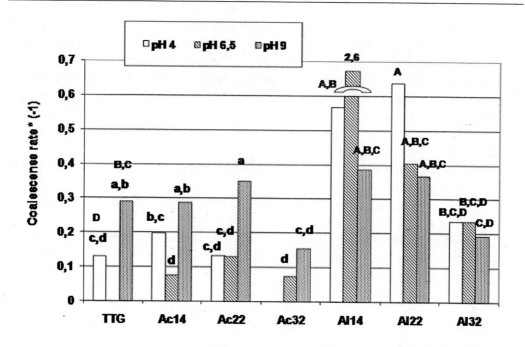

Figure 12. Coalescence rate from emulsions made with different extracts at pH 4, 6.5 y 9 for thermal treated gluten (TTG) and hydrolyzates. Lowercase letters: ANOVA for emulsions made with TTG and Ac hydrolyzate extracts; capital letters: ANOVA for emulsions made with TTG and Al hydrolyzate extracts. Different letters mean significant differences among samples (p<0.05).

ANALYSIS OF EMULSIONS COALESCENCE

The coalescence rates are negative because the curves are decreasing (negative ΔBS) (Figure 12). The values of Al14 at pH 6.5 were not considered for ANOVA because of the high value of instability.

In general, emulsions made whit Ac hydrolyzate extracts were more stable to the coalescence than those made with Al hydrolyzate extracts.

For TTG and Ac emulsions, it was observed that coalescence was higher at pH 9 than at pH 4. At this pH Coalescence Rate was lower at higher DH (Ac32 did not present coalescence). Also, at pH 9 coalescence rate was lower than at higher DH (32%).

The coalescence of emulsions made whit extracts at pH 6.5 was independent from DH and lower than that at pH 9. Thus for Ac-hydrolyzates extract emulsions, coalescence was higher at pH 9 and lower at pH 6.5 and 4.

Al-emulsions had higher coalescence rates than the emulsions made with the same DH for Ac, at all the pH evaluated. Always was observed that coalescence rate was lower at higher DH. This could be related with the higher content of medium hydrophobicity components presented in Al32 hydolyzates observed in these samples by Drago *et al.* (2008b).

The coalescence depends on the rupture of the interfacial film, which depends on viscoelastic properties and film mechanical resistance.

The effect of pH could be explained according to Phillips (1981). The rupture of the films and the coalescence of droplets could take place when repulsion forces (electrostatics and steric repulsions) are not high enough to compensate attractive forces (London and van der Waals). At higher steric repulsion, higher is the thickness of interfacial film.

The presence of amphiphilic peptides with charges is important for avoiding droplet coalescence (Caessens et al. 1999). However, the sign of the charge could play a particular role. At pH 9, the film formed by adsorbed proteins is negatively charged. Vojdani and Whitaker (1994) observed for hydrolyzates of β-lactoglobulin with Glu-C enzyme, a better emulsion capacity and stability than native protein, due to negatively charge groups from glutamic acid.

However, results showed that the effect of pH depended on the kind of hydrolyzates (DH, enzyme). Regards this, emulsions made at pH 9 were more stable for Al extracts and the contrary was for Ac extracts. Also, for Ac14 and Ac22 at pH 6.5 a lower coalescence rate than at pH 9 was observed. In this case, at pI peptides could have a more compact structure and form cohesive film by hydrophobic interactions (Cheftel et al. 1989), avoiding unfolding, peptide desorption of the film and lamella rupture that lead to coalescence.

CONCLUSIONS

Regarding the influence of DH and pH, no significant differences were observed for emulsion capacity of the different extracts of Al and Ac. The different behavior of Ac-6.5 extracts is related with the tendency to flocculation and not with a different ability to form the emulsion

Regarding stability, emulsions made with TTG extracts at pH 4 and 9 presented similar creaming rate, but coalescence was lower at pH 4 than 9.

For emulsions made whit Ac extracts, at pH 4, creaming rate was higher at the highest DH. However, Ac32-4 presented lower coalescence than the samples with lower DH or without hydrolysis. At pH 9, DH increase creaming rate, but coalescence rate was lower for Ac32. At pH 6.5 a creaming of a flocculated emulsion was observed. Thus, different mechanisms are involved in instability: creaming at pH 4 and 9 and strong flocculation at pH 6.5. Moreover, coalescence was higher at pH 9 and lower at pH 6.5 and 4. For Ac extracts, higher DH was related with lower creaming stability.

In general, emulsions made with Al extracts were more creaming-stable than Ac-emulsions and had little differences among their emulsions.

Emulsions made with Al-extracts showed higher coalescence rates than the respectively DH-Ac-emulsions at all evaluated pH. Moreover, at the evaluated pH, coalescence rate decreased at higher DH. This is related with the nature of the peptides produced by hydrolysis.

The effects of DH and pH on stability were different according to the enzyme used for hydrolysis, being emulsions made with Al extracts more stable to the creaming and less stable to the coalescence than emulsions made with Ac extracts.

Both, charge and peptide composition depend on pH extraction and influence the hydrophilic/hydrophobic balance of the peptides. Fractionating by pH gives peptide extracts with similar emulsion capacity but with different emulsion stability.

Acknowledgment

Partially support by Proy. CAI+D 2009 - II PI -54-258

References

AACC (1983) *Approved Methods of the American Association of Cereal Chemists. 8th Edition.*

Abismail B., Canselier J.P., Wilhelm A.M., Delmas H., Gourdon C. (1999) Emulsification by ultrasound: drop size distribution and stability. *Ultrasonics Sonochemistry* 6, 75–83

Abismail B., Canselier J.P., Wilhem A.M., Delmas H., Gourdon C. (2000) Emulsification processes: on line study by multiple light scattering measurements. *Ultasonics Sonochemistry*. 7, 187-192

Agboola, S.O., Singh, H., Munro, P.A., Dalgleish, D.G., and Singh, A.M. (1998) Desestabilization of oil in water emulsions formed using highly hydrolyzed whey proteins. *J Agric Food Chem.* 46, 84-90

Anton, M., Beaumal, V., Brossard, C., Llamas, G. and Le Denmat, M. (2002) Droplet flocculation and physical stability of oil-in water emulsions prepared with hen egg yolk. In: *Food emulsions and dispersions*. Ed. Anton M, Research Signpost, Kerala, India. pp. 15-28

Babiker, E. F., Fujisawa, N., Matsudomi, N., and Kato, A. (1996). Improvement in the functional properties of gluten by protease digestion or acid hydrolysis followed by microbial transglutaminase treatment. *J. Agric. Food Chem.* 44, 3746–3750.

Caessens, P.W.J.R., Gruppen, H., Slangen. C.J., Visser. S., and Voragen. A.G.J. (1999) Functionality of β-casein peptides: importance of amphipathicity for emulsion-stabilizing properties. *J. Agric. Food Chem.* 47, 1856-1862

Cheftel, J.C., Cuq, J.L. y Lorient, D. (1989) Propiedades funcionales de las proteínas. In: *Proteínas Alimentarias*. Ed. Acribia S.A. Zaragoza, España. Cap. 4, pp. 49-105

Dalgleish, D.G. (2006) Food emulsions- their structures and structures-forming properties. *Food Hydrocolloids*. 20, 415-422

Damodaran, S. (1989) Interrelationship of Molecular and Functional Properties of Food Proteins. En: *Food Proteins*. Ed. Kinsella, J.E. and Soucie, W.G. The American Oil Chemist' Society. cap.3, 21-51

Damodaran, S. (1994) Structure-Function Relationship of Food Proteins. In: *Protein Functionality in Food Systems*. Ed. Hettiarachchy, N.S. Ziegler G.R. IFT Basic Symposium Series 9, Chicago, Illinois. Cap. 1, pp. 1-37

Dickinson, E. (1998) Structure, stability and rheology of flocculated emulsions. *Current Opinion in Colloid & Interface Science.* 3: 633-638

Drago, S.R., and González, R. (2001). Foaming properties of enzymatically hydrolyzed wheat gluten. *Innovative Food Science & Emerging Technologies*, 1, 269-273.

Drago, S.R., González R.J. and Añón M.C. (2008a) Application of surface response methodology to optimize hydrolysis of wheat gluten and characterization of selected hydrolysate fractions. *J Sci. Food and Agric.* 88 1415-1422 (2008).

Drago, S.R., González, R.J. and Añón, M.C. (2008b) Techno-functional properties from hydrolyzed wheat gluten fractions. In: *Food Science and Technology: New Research.* Editors: L. V. Greco and M. N. Bruno cap. 10, pp. 355-381. 2008 Nova Science Publishers, Inc. Hauppauge, NY

Elizalde B.E., Pilosof, A.M.R., Bartholomai, G.B. (1991) Prediction of Emulsion Inestability from Emulsion Composition and Physicochemical Properties of Proteins. *J. Food Sc.* (56) 1: 116-120

Halling, P.J. (1981) Protein-stabilized foams and emulsions. *CRC Critical Reviews Food Sci. and Nutr.* 15, 155-203

Kato, A. and Nakai, S. (1980) Hydrophobicity Determined by a Fluorescence Probe Method and its Correlation with Surface Properties of Proteins. *Biochim. Biophys. Acta* 624, 13-20

Kato, A., Osako, Y., Matsudomi, N., and Kobayashi, K. (1983) Changes in the emulsifying and foaming properties of proteins during heat denaturation. *Agric. Biol. Chem.* 47, 33-38

Kempa, L., Schuchmann, H.P., and Schubert, H (2005) *The role of surfactant in emulsification.* Proceedings Intrafood- EFFoST Conference. Ed. P. Fito and F. Toldrá, 25-28 octubre, Valencia, España. (1): 759-762

Kong, X., Zhou H., and Qian H (2007) Enzymatic preparation and functional properties of wheat gluten hydrolyzates. *Food Chemistry* 101, 615–620

Lemarchand C., Couvreur P., Vauthier C., Costantini D., Gref R. (2003) Study of emulsion stabilization by graft copolymers using the optical analyzer Turbiscan. *Int. J Pharm.* 254, 77–82

Linares, E., Larré, C., Lemestre, M., and Popineau, Y. (2000) Emulsifying and foaming properties of gluten hydrolysates with an increasing degree of hydrolysis: role of soluble and insoluble fractions. *Cereal Chem.* 77(4) 414-420

Linares E, Larre C, Popineau Y. (2001a) Freeze- or spray-dried gluten hydrolysates. 1. Biochemical and emulsifying properties as a function of drying process. *J Food Eng.* 48, 127-135

Linares E., Larre C., Popineau Y. (2001b) Freeze- or spray-dried gluten hydrolysates. Effect of emulsifcation process on droplet size and emulsion stability. *J Food Eng.* 48, 137-146

Pan, L.G., Tomás, M.C., and Añón, M.C. (2002) Effect of sunflower lecithins on the stability of water-in-oil and oil-in-water emulsions. *J Surfactants Detergents.* 5, (2): 135-143

Pan, L.G., Tomás, M.C., and Añón, M.C. (2004) Oil-in-water emulsions formulated with sunflower lecithins: vesicle formation and stability. *JAOCS.* 18, (3): 241-244

Pizzino A. Catté M, Van Hecke E, Salager JL, Aubry JM. (2009) On-line light backscattering tracking of the transitional phase inversion of emulsions. Colloids and Surfaces A: Physicochem. *Eng. Aspects* 338, 148–154

Phillips, M.C. (1981) Protein conformation at liquid interfaces and its roles en stabilizing emulsions and foams. *Food Tech.* 35 (1): 50-57

Phillips, R. D. and Beuchat, L. R. (1981). Enzyme modification of proteins. In: Cherry, J.P. (Ed.), Protein functionality in foods, (pp. 275–298). *ACS Symposium Series,* Washington, DC.

Popineau, Y., Huchet, B., Larré, C., and Bérot, S. (2002) Foaming and emulsifying properties of fractions of gluten peptides obtained by limited enzymic hydrolysis and ultrafiltration. *J. Cereal Sci.* 35, 327-335

Quintana, M., Califano, A., and Zaritzky, N. (2002) Stability and rheological properties of food emulsions as affected by HLB, thickeners, salt and oil content. En: *Food emulsions and dispersions*. Ed. Anton M, Research Signpost, Kerala, India, pp 2-13

Rahali, V., Chobert, J.M., Haertle, T., and Gueguen, J. (2000) Emulsification of chemical and enzymatic hydrolysates of β-lactoglobulin: characterization of the peptides adsorbed at the interface. *Nahrung*. 44, 89-95.

Robins, M.M. (2000) Lipid emulsions. *Grasas y Aceites*. 51 (1): 26-34

Santiago, L.G., Bonaldo, A.G., and González, R.J. (2000) Formación y estabilidad de emulsiones de aislados proteicos de soja con diferente contenido de sulfhidrilos libres. *Food Sci. Tech. Int.* 6 (5) 407-414

Santiago, L.G., Bonaldo, A.G., and González, R.J. (2004) Influence of salt level and pH on the stability of oil-in-water emulsions containing untreated, heat and reduced treated soy protein isolate as emulsifier. *Braz. J. Food Tech.* 7 (2): 179-186

Spellman, D., McEvoy, E., O'Cuinn, G., & Fitz Gerald, R. (2002). Proteinase and exopeptidase hydrolysis of whey protein: Comparison of the TNBS, OPA and pH stat methods for quantification of degree of hydrolysis. *Int. Dairy J.* 13, 447-453.

Stainby, G. (1986). Foaming and Emulsification. In: *Functional Properties of Food Macromolecules*. Mitchel, J.R.; Ledward D.A. Elsevier Applied Sci. Publishers Ltd. cap 7, pp. 315-353

Tadros, T.F. and Vincent, B. (1983). Emulsion stability. In: Becher P, editor. *Encyclopedia of Emulsion Technology*. Vol. 1. New York: Marcel Dekker Inc. pp 129-285.

Tornberg, E; Olsson, D. and Pearson, K. (1990). The Structural and Interfacial Properties of Food Proteins in Relation to their Function in emulsions. En *Food Emulsions*. Larson, D.; Friberg S., Ed. Marcel Dekker, New York, USA

Van der Ven, C., Gruppen,H., de Bont, D.B. A., and Voragen AGJ. (2001) Emulsion Properties of Casein and Whey Protein Hydrolysates and the Relation with Other Hydrolysate Characteristics *J. Agric. Food Chem.* 49, 5005-5012

Vioque, J., Clemente, A., Pedroche, J., Yust, M., & Millán, F. (2001). Obtención y aplicaciones de hidrolizados proteicos. *Grasas y Aceites*. 52, (2): 127-131.

Vojdani, F. and Whitaker, J.R. (1994) Chemical and enzymatic modification of proteins for improved functionality. In: *Protein Functionality in Food Systems*. Ed. Hettiarachchy, N.S. and Ziegler, G.Z. Marcel Dekker, Inc. Chicago, Illinois. Cap. 9, pp. 261-309

Voutsinas, L.P., Cheung, E. and Nakai, S. (1983) Relationships of hydrophobicity to emulsifying properties of heat denatured proteins. *J. Food Sci.* 48, 26-32.

Chapter 8

TECHNO-FUNCTIONAL PROPERTIES FROM HYDROLYZED WHEAT GLUTEN FRACTIONS[*]

S. R. Drago[1], R. J. González[1] and M. C. Añón[2]

[1] CONICET, Inst. de Tecnología de Alimentos -FIQ- Univ. Nacional del Litoral,
1 de Mayo 3250, Santa Fe, Argentina
[2] CONICET, CIDCA, calle 47 y 116, La Plata, Argentina

ABSTRACT

Many protein sources that are found in the market are obtained as by-products and there is a great interest in using them as protein ingredients with adequate functionality for food formulation. Structure modification permits one to add value and to diversify their uses. The diversification of wheat gluten applications depends on the improvement of its solubility in a wider pH range. One of the alternatives that allow protein modification in these products is the enzymic hydrolysis. The objective of this work was to evaluate foaming properties of protein fractions obtained by extracting, at 3 pH, different hydrolyzed gluten samples. Two commercial enzymes (acid -Ac- and alkaline -Al- proteases) were used to reach 3 different hydrolysis degrees (DH). Extracts pHs (4, 6.5 and 9) were diluted to a protein concentration of 4 g/l. RP-HPLC, free amino groups content, sulphydryl and disulfur content, average peptide chain length were used to characterize each extract. Foam was produced by sparging nitrogen at a known rate through a dilute protein solution. The maximal volume of liquid incorporated into the foam (Vmax) and the rate of liquid incorporation into the foam (Ri) were determined and used as indicators of foaming capacity. The times for half-drainage of the liquid that was incorporated into the foam at the end of the bubbling period ($t_{1/2}$) and the rate of liquid drainage from the foam were also measured. Regarding Ri, all pH 4 extracts from hydrolyzed samples showed higher Ri than an un-hydrolyzed sample. Extracts from Al hydrolyzed extracts showed higher Ri than those from Ac. In the case of Al extracts, an inverse relation between DH and Ri was observed, but practically no influence of DH on Ri, was observed in the case of Ac extracts. For pH 6.5 extracts, the relation between DH

---

[*] Chapter 8 – A version of this chapter was also published in *Food Science and Technology: New Research*, edited by Lorenzo V. Greco and Marco N. Bruno, published by Nova Science Publishers, Inc. It was submitted for appropriate modifications in an effort to encourage wider dissemination of research.

and Ri were in opposite directions, depending on the enzyme, for Ac, Ri decreased with DH, while for Al, Ri increased with DH. At this pH, it was observed that the extracts which foamed more quickly, were those with the highest times for half-drainage of the liquid ($t_{1/2}$). Some foam parameters correlated between themselves, depending on the extracts. Foaming capacity and stability depend on pH, DH and enzyme and it was possible to correlate parameters with composition evaluated by RP-HPLC.

INTRODUCTION

Many protein sources that are found in the market are obtained as by-products and there is a great interest in using them as protein ingredients with adequate functionality for food formulation. Wheat gluten is a by-product of the wheat starch process, which is available in large amounts and at relatively low cost. Because it is insoluble in water at near-neutral pH and is viscoelastic when hydrated, gluten is mainly used to enhance the strength of flours for breadmaking and for textured food (Popineau et al. 2002; Babiker et al. 1996). Enzymic hydrolysis is one of the alternatives that allow protein modification by means of structural transformations, which could provide other properties to it (techno-functional or bio-functional properties).

Foaming capacity is a techno-functional property that allows protein ingredients to be used to improve texture, consistency and appearance of foods (Kokini and Aken, 2006).

The behavior of proteins at the air/water interface has been described by many authors (Mac Ritchie, 1978; Graham and Phillips, 1979a,b,c; Kinsella, 1981; Kinsella and Phillips, 1989; Damodaran, 1994). The ability of proteins to act as surfactants and stabilize foams and emulsions depends upon their ability to adsorb at interfaces, greatly reduce the interfacial tension, and form a cohesive film. Since all proteins are amphiphilic, they show a tendency to adsorb at interfaces (Tsaliki et al., 2004). However, the foam ability differs widely among proteins. These differences arise primarily from differences in conformation and in the physicochemical properties of the protein surfaces that interacts with the dispersed and continuous phases in the foam (Damodaran, 1994).

The foaming capacity of a protein depends on many events. For surface adsorption, proteins must follow successive steps: a) diffusion from the bulk solution to the interface, (which kinetic depends essentially on the size of the hydrated molecule) overcoming the electrostatic barrier to the adsorption process (the higher the charge of the molecule, the higher the energy required); b) adsorption (which depends mainly on the proportion of non polar residues, on their repartition along the chain and on the flexibility of the macromolecule; c) the conformational rearrangements for surface coverage and loop and tail formation are easier when molecules are flexible; d) film formation at the interface whose thickness and resistance depend on the tertiary structure of the protein and on the protein-protein interactions (Le Meste et al., 1990).

Also, protein surface properties are related to the presence of components with adequate surface hydrophobicity (Nakai, 1983).

Moreover, it is known that the pH solution affects foam properties, modifying the protein net charge and the formation and properties (thickness and viscoelaticity) of the protein film (German and Phillips, 1994).

It has been observed that the mild protein hydrolysis increase foaming and emulsifiying properties but impair the stability (Linares et al., 2000).

Ultra-filtration has been used to fractionate protein hydrolysates in order to select fractions with good foaming properties (Popineau et al., 2002). Another alternative would be the fractionation by extraction at different pH.

The aim of the present work was to evaluate foaming properties of peptide fractions, obtained by extracting at 3 pH different hydrolyzed gluten samples with different degree of hydrolysis and from two proteolytic enzymes, and to correlate these properties with their structural characteristics.

MATERIAL AND METHODS

Commercial vital gluten provided by Molinos SEMINO S.A. (Carcarañá - Santa Fe) was used. Gluten composition was the following: moisture: 5.95% (AACC 44-15A method)[10], protein (N x 5.7): 77.20 % d.b. (Kjeldahl –AACC 46-11 method), starch: 13.15 % d.b. (Ewers polarimetric method), ether extract 0.71 % d.b. (AACC 30-25 method) and ash: 0.834 % d.b. (ICC N° 104- IRAM N° 15851 Standard technique). In order to disperse vital gluten in water and secure a uniform suspension, a moderate thermal treatment was carried out (Drago et al, 2008). The thermal-treated gluten (TTG) sample was used in all experiments as a substrate. The acid and alkaline enzymes used in the experiment were provided by GENENCOR S.A. (Arroyito-Córdoba). The acid enzyme (Ac) was a fungal protease derived from *Aspergillus oryzae* (31.000 HU/g), which is a mixture of endo/exopeptidases whose characteristics are the following: effective pH: 3.5 – 9, optimum pH: 4.3 – 5, temperature range: 30 – 50 °C. The alkaline enzyme (Al) was an endoprotease derived from *Bacillus Licheniformis*, which effective pH rangr was 7- 10, optimum pH: 9.5, temperature range: 25-70 °C, optimum: 60°C.

Preparation of Hydrolysates

Hydrolysates were prepared in a 5 l batch reactor with agitation, using a thermostatized bath. Protein concentration was 8% (W/W) and 3N HCl or NaOH were added in order to reach and maintain a constant pH. After enzyme inactivation, the hydrolysates were frozen at -20 °C and lyophilized. Reaction parameters used for acid protease were: temperature: 55°C, pH: 4.25, Enzyme/Substrate ratio (E/S): 5%. Enzyme inactivation was carried out at 70°C for 15 min. Reaction parameters used for alkaline protease were temperature: 60°C, pH: 9.5, E/S ratio: 0.095%. Enzyme inactivation was carried out at 80-85°C for 10 min.

Hydrolysates were obtained at three different trichloroacetic acid index (TCAI) for both acid and alkaline enzyme hydrolysates: 14, 22 and 32%, named Ac14, Ac22, Ac32 for acidase enzymic hydrolysates and Al14, Al22, Al32 for alkaline enzyme hydrolysates.

Hydrolysis Reaction Progress

The hydrolysis progress was followed by means of the trichloroacetic acid index (TCAI), using TCA 20% and diluting the sample in a 1:1 ratio. N was measured by the Semimicro-Kjeldahl method. The TCAI, which was used as an indirect measurement of degree of hydrolysis (DH), was calculated as follows:

TCAI = [N soluble in TCA (hydrolyzate) − N soluble in TCA (blank) x 100] / total aminic N

Preparation of Fractions at Different pH

In order to obtain the hydrolysates fractions at different pH (4, 6.5 and 9), a 2% (W/W, dry basis) solution of the different hydrolysates was prepared (Drago and González, 2001). The pH was achieved by adding 0.8N HCl or 0.8N NaOH. The samples were stirred for 1 hour at room temperature, and then centrifuged during 15 min at 8000xg at room temperature. The supernatant (the extract at each pH) was frozen and protein content determined by semimicro-Kjeldahl method.

The solubility was calculated as Nsupernatant (g) x 100/N total (g)

Characterization of Soluble Fractions at Different pH of Thermally Treated Gluten (TTG) and Hydrolysates

Reverse Phase HPLC (RP-HPLC) of the Extracts

The extracts were diluted to a protein concentration (Nx5.7) of 2.5 mg/ml, and a Sephasil Peptide C_8 column of 12 µm ST 4.6/250 (Pharmacia Biotech) was used, together with an auto injector Waters 717 Plus Autosampler − Millipore, a Waters 600 E pump (Multisolvent Delivery System- Millipore), and a diode array detector (Waters 996-Millipore). Peptides were separated and eluted at 1.1 ml/min, at 60°C, by using the following buffers: Buffer A: Acetonitrile-water 2:98, with 650 µl/l of trifluoroacetic acid (TFA); Buffer B: Acetonitrile-water 65:35 with 650 µl/l of TFA, and detected at 210 nm. Since elution profiles of RP-HPLC can be grouped in categories according to the increasing hydrophobicity of the eluted peptides (Linares et a.l, 2001), chromatogram analysis was carried out by integrating peak areas in three sections of each chromatogram:

   a. Components of low molecular weight and low hydrophobicity: 0-20 min of elution range (LH)
   b. Components of medium molecular weight and medium hydrophobicity: 20-40 min of elution range (MH)
   c. Components of high molecular weight and high hydrophobicity: 40-60 min of elution range (HH)

Results were expressed as a percentage of each section with respect to the total area.

Determination of Free Amine Group Content

The o-phthaldialdehyde (OPA) technique (Nielsen *et al.*, 2001) was used for this purpose. Free amine group content was used to calculate the number of peptide bonds cleaved during hydrolysis.

Estimation of Peptide Chain Length (PCL)

PCL can be estimated as was decrypted by Adler Nissen (1986) by means of the following expression:

PCL (soluble fraction) = 100 x %S / (h/ h_{tot})%.

Where:

h_{tot}, is the total number of peptide bonds in the protein substrate (8.3 mEq/g protein)
h, is the number of peptide bonds cleaved during hydrolysis.
%S: fraction of soluble proteins.

Determination of Thiol Groups and Disulphide Bonds

Thiol groups (SH) and disulphide bonds (SS) in hydrolyzed gluten fractions were determined colorimetrically with 5,5'-dithiobis (2-nitrobenzoic acid) (DTNB) by means of a Anderson y Wetlaufer (1975) method, modified by Graveland *et al.* (1985).

For the determination of thiol groups, samples were dissolved in 1 ml of the following stock solution: 6.0 M urea, 0.05M SDS, 0.1 M sodium phosphate buffer pH 7.0, and 1 mM EDTA. A 0.1 ml aliquot of 0.01 M DTNB, dissolved in the same stock solution, was then added, and after 5 min the absorbance was measured at 412 nm.

Disulphide bonds were estimated after alkaline cleavage. A 1 ml sample of the hydrolysate fraction, in the same solution used for thiol determination, was mixed with 1 ml 6.0 M NaOH and stirred for 30 min at 50°C. The reaction mixture was neutralized by adding 2.0 M H_3PO_4 (1.0 ml), mixed thoroughly and the chromophore was developed by adding 0.01 M DTNB solution (0.1 ml). Bovine serum albumen was used as standard (Kella and Kinsella, 1985). Disulphide bonds were estimated using the following:

SS = ½ (SH total - SH free) (μmol)

Foaming Properties

Foaming properties of fractions obtained at pH 4, 6.5 and 9, from gluten hydrolyzed with acid or alkaline enzymes and from TTG were evaluated. All determinations were performed four times. N_2 gas was sparged during 1 min at a flow rate of 40 ml/s, through 6 ml of 4 g/l of protein sample, contained into a 3x30 cm glass column. During the test, the conductivity was recorded along the bubbling time and during 10 min after stop the sparging. Volume of liquid in the foam was estimated according to Popineau *et a.l*, (2002). Conductivity measurements as a function of time (Ct) and with reference to the conductivity of test solution (Ci) were used to calculate the volume of liquid in the foam (V_L):

V_L (ml) : V_0 [1- (Ct /Ci)]

Where:

V_0: is the volume of sample solution (6 ml) introduced into the sparging chamber

The maximal volume of liquid incorporated to the foam (Vmax, ml), the liquid volume in the foam 10 min after the end of the bubbling (Vf, mL), and the rate of liquid incorporation to the foam (Ri, ml/min) were determined. Ri is related to the surface behavior of the protein at the initial step of foam formation (Wagner *et al.*, 1996).

Foam stability was estimated from the time for half-drainage of the liquid that was incorporated into the foam at the end of the bubbling period ($t_{1/2}$, min). The initial rate of liquid drainage was also measured. The decay portion of the curve: volume of liquid incorporated to the foam (ml) vs time (min), was fitted with a single exponential decay.

$$y = y0 + A \, e^{-k.t} \tag{1}$$

Where,

y: is the liquid into the foam (ml).
A and k: were equation parameters.
The first derivative of (1) respect to the time, at t=0 is the initial rate of liquid drainage:
dy/dt = -A . k = V_0dren
Figure 1 shows an example of the record of the formation and the instability of the foam.

Statistical Analysis

Software Statgraphics Plus 3.0 was used for statistical analysis. LSD (Least significant difference) test was used to determine statistical differences among samples ($p<0.05$).

RESULTS AND DISCUSSION

Thermal Treated Gluten (TTG) Hydrolysis

Figure 2 shows that TCA solubility increase with time for both enzymes. From 20% TCAI, higher rate is observed for Al and 32% TCAI is reached faster. This level was the maximum hydrolysis limit for both enzymes, and it was selected in order to avoid impairing surface properties by an excessive hydrolysis.

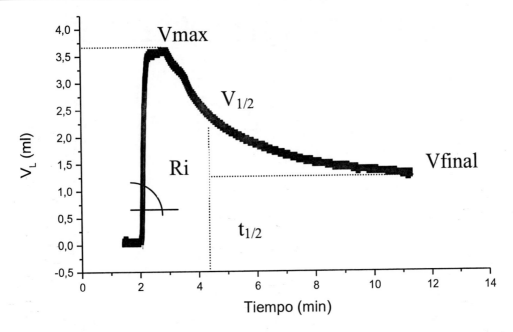

Figure 1. Foam formation and unstabilization. Ri: rate of liquid incorporation to the foam; Vmax: maximal volume of liquid incorporated to the foam; $V_{1/2}$: foam volume at $t_{1/2}$; $t_{1/2}$: time for the half-drainage of the liquid, Vfinal: final volume 10 min after bubbling stops.

Solubility and Reverse Phase HPLC (RP-HPLC) of the Extracts

Figure 3 shows the protein solubility corresponding to thermally treated gluten and the six hydrolysates. Hydrolysate solubility depended on pH, DH, and the physicochemical characteristic of peptides and consequently depended on the site of peptide cleavage, which change with the enzyme used (Chobert et al., 1996).

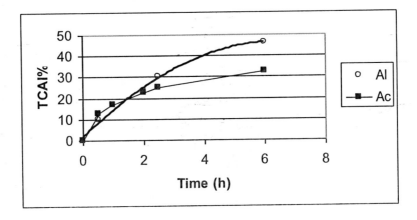

Figure 2. Hydrolysis curve. Ac: acid protease. Substrate concentration: [S]= 8%, pH=4.25, T=55°C, E/S= 5%; Al: Alkaline protease: [S]= 8%, pH=9.5, T=60°C, E/S= 0.095%.

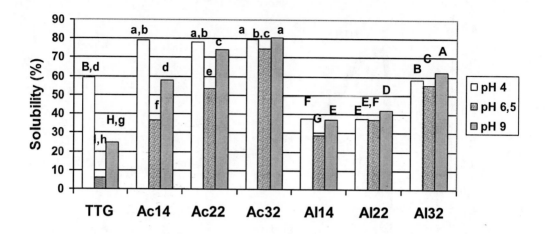

Figure 3. Solubility percentage of thermal treated gluten (TTG) and for Ac (acid protease) and Al (Alkaline protease) hydrolyzates. Different letters means significant differences among samples (p<0.05).

Table 1. Integrated areas of RP-HPLC chromatograms (percentage of total area)

Sample	pH	LH 0-20 min	MH 20-40 min	HH 40-60 min
TTG	4	8,71	25,11	66,14
	9	11,18	32,58	56,36
Al 14	4	8,73	57,43	32,29
	6.5	13,26	55,84	30,83
	9	11,70	56,54	31,93
Al 22	4	11,43	57,98	30,52
	6.5	10,59	62,17	26,79
	9	9,87	60,16	31,05
Al 32	4	10,27	64,56	26,13
	6.5	11,76	60.14	28,08
	9	12,33	60,89	26,76
Ac 14	4	11,15	42,24	46,54
	6.5	14,28	52,35	33,29
	9	12,38	46,21	41,33
Ac 22	4	12,55	47,69	40,13
	6.5	14,25	52,71	32,97
	9	12,44	50,05	38,18
Ac 32	4	13,60	58,13	29,67
	6.5	19,43	42,91	37,19
	9	13,40	53,92	33,51

HH: high hydrophobic and high molecular weight components; MH: medium hydrophobic and medium molecular weight components; LH: low hydrophobic and low molecular weight components.

Table 1 shows the reverse phase HPLC (RP-HPLC) of the extracts. The percentage of each section (LH, MH and HH) is the ratio of the corresponding integrated areas with respect to the total area of chromatogram. For small peptides (<15 residues), the hydrophobicity of the amino acid side chains in the peptides determine the elution time from a reversed phase column. However, for larger peptides, the peptide length also influences retention time (van der Ven, 2001).

Although hydrolysis increased gluten solubility, as it is generally found with protein hydrolysis (Linares et al., 2000), at the same TCAI the behavior of hydrolysates was different according to the enzyme used.

At pH 4, Ac hydrolysate had higher solubility than that of TTG, but subsequent hydrolysis did not cause further increase. This could be attributed to the fact that the enzyme cleaved the available HH soluble components at acid pH. Table 1 shows that at pH 4 as TCAI increase MH components of Ac extracts increase as a consequence of the decrease of HH components. In the case of Al hydrolysates it is observed that they had lower solubility than those of TTG and reach equal value at TCAI 32%. This could be related with the fact that hydrolysis was carried out at alkaline pH, where gluten has low solubility (aproximately 25%), and the enzyme first cleaved the available soluble polypeptides and then hydrolyzed more hydrophocic and insoluble polipeptides. As it is shown in Table 1, acid extract of TTG have more content of HH components than pH 9 extract. Besides that new peptides obtained by hydrolysis had also lower solubility at pH 4 than the non hydrolysed protein probably due to aggregation fenomena during alkaline hydrolysis.

At pH 6.5 (isoelectric point region) and 9, solubility from hydrolysates was higher than that of TTG and also, for both enzyme hydrolysates, higher solubility is obtained as TCAI increases.

For Ac, at pH 6.5 (at the isoelectric point region) hydrolysate solubility was always significantly lower than that corresponding to other pH. Extracts at this pH have higher LH component content than the other pH extracts. The 14 and 22 TCAI extracts are poorer in high hydrophobic components because of they are insoluble at the pI. This could be explained taking into account the way the Ac enzyme cleaves the protein molecules, producing big size polypeptides, whose solubility is more pH dependable, together with small soluble peptides.

For Al hydrolysates, pH affect solubility in a lesser extension than Ac hydrolysates. However, for the same TCAI, Al hydrolysates had always lower solubility than that of Ac hydrolysates and it was observed that as DH increase MH component content increase and HH decrease.

At pH 9, for Ac extracts, as TCAI increases, content of MH component increases but that of HH decreases, while for Al extracts no clear effects were observed.

Extract Free Amino Group Content

Figure 4 shows the amount of free amino group per 100 g of soluble protein for TTG and the three different hydrolysates extracts of both enzymes. In general, as protein is hydrolyzed amino group content increases and a mixture of peptides having different molecular weight is obtained. For Ac14 and Ac22 extracts at pH 6.5 (near isoelectric point), components having low molecular weight predominate because of the solubility of high molecular weight components is very low (as a consequence of their reduced charge).

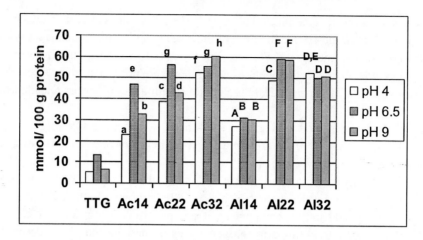

Figure 4. Free amino groups/100g of protein for thermal treated gluten (TTG) and for Ac (acid protease) and Al (Alkaline protease) hydrolyzates. Different letters means significant differences among samples (p<0.05).

For Ac32 extracts, there are more components of low molecular weight and the impact of pH is less important. This is also observed with PCL (Figure 4).

For Al extracts, the behaviour is different. It is observed that the proportion of amine group with respect to protein is similar for the extracts of the same TCAI hydrolysate.

Average Peptide Chain Length

Figure 5 shows the average peptide chain length of the soluble fraction vs. TCAI at each pH for hydrolysates of both enzymes (Al and Ac).

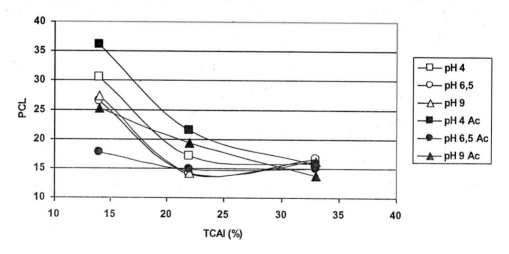

Figure 5. Relationship between medium peptide chain length (PCL) and TCAI %, for extracts obtained with Ac and Al enzymes. Different letters implies significant differences among samples (p<0.05).

Taking into account that TTG PCL is 162, 62 and 131 at pH 4, 6.5 and 9, respectively, at 14% TCAI an important decrease in the size of polypeptides of soluble fractions is observed. For Ac14 and Ac22, PCL is significantly affected by the pH but for Al hydrolyzates, PCL is practically the same for the same TCAI, independent of the pH.

For TCAI 14 hydrolysates of both enzymes, at pH 4 higher size components were extracted in comparison with other pH extracts.

As it was observed with free amino group content, the extract at pH 6.5 corresponding to Ac and for the same TCAI, has the peptides with the smallest sizes. For the hydrolysates with 32%TCAI, protein extractability was not much affected by pH, and their extracts have peptides of about 14 amino acid residues.

At the first steps of the hydrolysis (TCAI 14 and 22%), Ac protease cleaved proteins giving high MW components (with different solubility at different pH, Figure 3), and low MW components. On the other hand, Al protease gave peptides of similar MW, and because of this, PCL of 22% has almost the same value than at 32% TCAI.

The way the enzyme acts can be seen with the PCL measured at the pH corresponding to the isoelectric point (near pH 6.5) or pH 9, where the solubility of the gluten thermally treated is low. PCL shows a rapid decrease with the progress of the enzymatic reaction during the first 30 min. This increase is then slower, which shows that soluble peptides are degraded during hydrolysis until they reach a certain length. After that, the enzyme would not act on them but on the substrates of higher MW, which increases solubility.

Thiol Groups and Disulphide Bonds Content

Figures 6 and 7 show thiol groups (SH) and disulphide bonds content (SS) respectively, of the extracts at different pH from the hydrolyzates obtained using both enzymes and TTG.

In general, the SS content of TTG is higher than those of the hydrolyzate extracts and Ac extracts present higher values than Al extract ones.

Al Extracts at pH 9 and 6.5 have higher SH content than Ac, while pH 4 Ac extracts have higher SH content than Al extract (except at 32% TCAI).

Soluble fractions from GTT are rich in SS. The values obtained are similar at those shown by Kasarda (Beveridge et al, 1974). Ac hydrolysates obtained at pH 4.25, have a lower SS content than that of TTG, but is still high. However, when Al is utilized, solubilized fractions have low content of SS but are rich in SH. This could be explained taking into account that at pH 9.5 the rupture and reduction of intermolecular SS take place. The pK of -SH group is about 8, and at pH higher than 8 interchange of thiol groups reactions could occur (Mutilangi et al., 1996). It is evident that these enzymes cleaved molecules in different way producing peptides having different composition and capacity to interact with other peptides. In this way, Al soluble fractions are poor in SS peptides because of SS rich peptides interact and precipitate, which is in accordance with the lower solubility of Al hydrolystes.

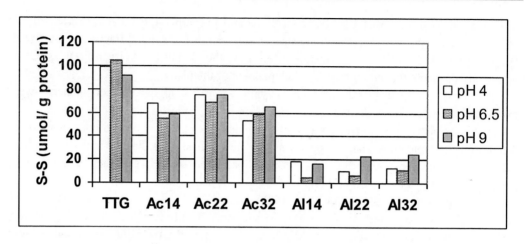

Figure 6. SS content in fractions obtained from: TTG (thermal treated gluten); Ac (acid protease) and Al (alkaline protease) hydrolyzates with different TCAI % (14, 22 and 32).

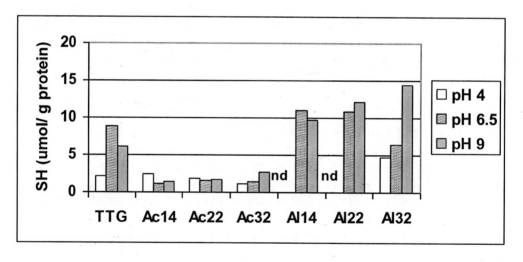

Figure 7. SH content in fractions obtained from: TTG (thermal treated gluten); Ac (acid protease) and Al (alkaline protease) hydrolyzates with different TCAI % (14, 22 and 32). nd: not detected.

Relationship Among Integrated Areas of RP-HPLC Chromatograms (HH, MH, LH) and PCL from Extracts

Some correlations between PCL and the integrated areas of RP-HPLC chromatograms were found. At both, pH 4 and 9, PCL is inversely related with the area of MH component (r: 0.771 and 0.7585, respectively), while it is directly related with the area of HH ones (r: 0.8347 and 0.798, respectively). These results suggest that when smaller components are produced (lower PCL), a reduction of product hydrophobicity takes place.

Evaluation of Foaming Capacity and Stability

Except for pH 6.5 TTG extract, which did not reach the desired protein concentration, foaming properties were evaluated for all extracts using 4 mg protein/ml dispersions.

Foaming capacity was evaluated using two different parameters related with this property: the rate of liquid incorporation to the foam (Ri, ml/min), which is related to the surface behaviour of the protein at the initial step of foam formation (Wagner *et a.l,* 1996) and the maximal volume of liquid incorporated to the foam (Vmax, ml).

Analysis of the Rate of Liquid Incorporation to the Foam (Ri)

Figure 8 shows Ri for the different extracts. Because the hydrophilic/hydrophobic balance depend on the pH and the extract composition, the analysis is made for each pH separately.

a. Analysis of pH 4 Extracts

At pH 4 all hydrolysate extracts showed higher Ri than TTG extracts. Gluten is a protein which has a high content of hydrophobic amino acids (37.5%). When DH increases, an increase of hydrophilicity is achieved due to free amino and carboxilic groups produced by proteolysis and so the hydrophilic/hydrophobic peptide balance changes. Also small peptide molecules (produced by hydrolysis) probably adsorb more easily than aggregates, because of their smaller size, and the higher availability of their residues.

Higher Ri values were obtained for Al extracts than Ac extracts. Moreover, while Al extract showed a Ri decrease as TCAI% increases, practically no change was observed for Ac ones. These results suggest that a mild hydrolysis improves surface activity.

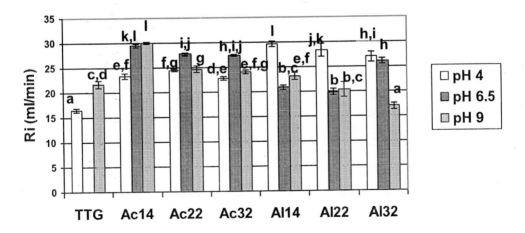

Figure 8. Rate of liquid incorporation to the foam (Ri) for different protein extract s (protein concentration: 4 mg/ml). Al (alkaline protease) and Ac (acid protease) for different TCAI %(14, 22 and 32).

When correlations between integrated areas of RP-HPLC chromatograms (HH, MH, LH) and Ri are analyzed, it is observed that MH content directly correlates with Ri at pH 4 (Figure 9) and that HH content inversely correlates with Ri (Figure 10).

This means that the presence of components with high hydrophobicity impairs foaming capacity and that Ac extracts contains more HH than Al extract.

We can say that the higher foam rate at pH 4 is related to peptides having medium molecular weight and medium hydrophobicity.

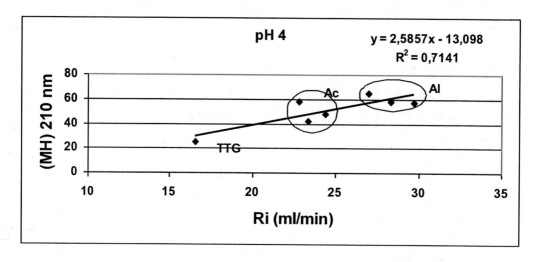

Figure 9. Relationship between rate of liquid incorporation to the foam (Ri) corresponding to extracts at pH 4 and MH (integrated areas of RP-HPLC chromatograms corresponding to medium hydrofobic and medium molecular weight components). Ac: acid protease. Al: alkaline protease.

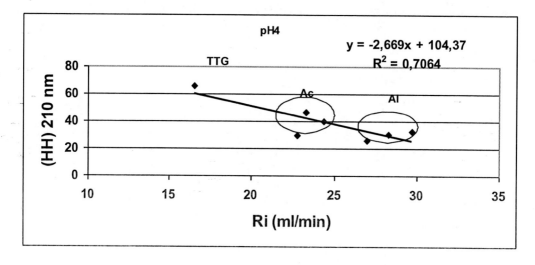

Figure 10. Relationship between rate of liquid incorporation to the foam (Ri) of extracts at pH 4 and HH (integrated areas of RP-HPLC chromatograms corresponding to high hydrophobic and high molecular weight components). Ac: acid protease. Al: alkaline protease.

b. Analysis at pH 6.5

Ac extracts at pH 6.5 showed higher Ri than at pH 4, and the contrary was for Al extracts. Moreover, DH affected Ri values in a different way, depending on the enzyme. For Ac extracts Ri decreases as DH increases, while for Al extracts Ri tends to increase.

At this pH no correlation between RI and integrated areas of RP-HPLC chromatograms was observed.

c. Analysis at pH 9

At pH 9, Ac extracts have higher Ri than TTG extracts but Ri decreases with the degree of hydrolysis. For Al extract the hydrolysis produced a decrease in Ri and only the sample with low TCAI had Ri higher than TTG.

For correlating integrated areas of RP-HPLC chromatograms with Ri, TTG extracts were discarded.

At this pH, a different behavior than pH 4 is observed. Ri increases with the presence of high hydrophoficity and high molecular weight components (HH) (Figure 11) and also as MH components decrease (Figure 12). However, it is important to point out that at pH 9 it is possible to extract peptides richer in SH and more flexible structures than at pH 4, and consequently it is possible to expect a high RI with a more hydrophobic- high molecular weight peptide.

Then, for a protein or peptide with good foaming capacity, measured as Ri, an adequate size, a ratio of hydrophilic/hydrophobic components and flexibility, are needed. In this way, TTG extracts with high MW, low flexibility and low hydrophilic/hydrophobic components ratio, showed lower Ri than hydrolyzed gluten extract with lower MW, higher flexibility and higher hydrophilic/hydrophobic components ratio, which make the adsorption at the interface easier.

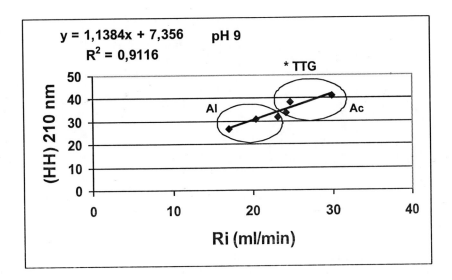

Figure 11. Relationship between rate of liquid incorporation to the foam (Ri) of extracts at pH 9 and HH (integrated areas of RP-HPLC chromatograms corresponding to high hydrophobic and high molecular weight components). Ac: acid protease. Al: alkaline protease.

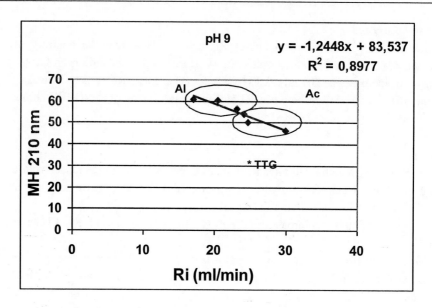

Figure 12. Relationship between rate of liquid incorporation to the foam (Ri) of extracts at pH 9 and MH (integrated areas of RP-HPLC chromatograms corresponding to medium hydrophobic and medium molecular weight components). Ac: acid protease. Al: alkaline protease.

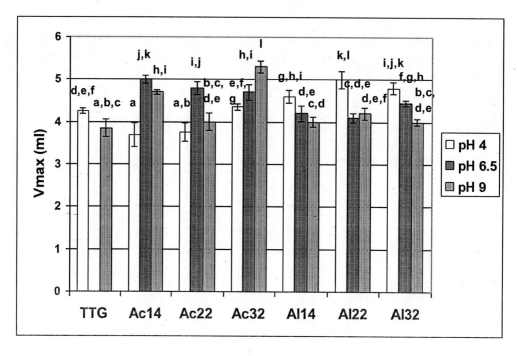

Figure 13. Maximal volume of liquid incorporated to the foam (Vmax.), for different hydrolyzate extracts (protein concentration 4 mg/ml) and TCAI % (14, 22 and 32). Ac: acid protease. Al: alkaline protease.

Analysis of Maximum Volume of Liquid Incorporated to the Foam (Vmax)

The Vmax of different extracts is shown in Figure 13. As it was observed, it is difficult to find a clear tendency between Vmax and TCAI% or pH.

The only correlation observed was between MH/HH and Vmax at pH 4 (Figure 14). At pH 4, higher MH components and lower HH, the peptides incorporated more liquid to the foam. This implies that a hydrolyzed structure is more capable of foaming than the raw protein.

Analysis of Foam Stability

Foam stability was measured through half-drainage of the liquid and the initial rate of liquid drainage.

Analysis of Half-drainage of the Liquid ($t_{1/2}$)

In general, foams made with Ac extracts were more stable than those of Al extracts (Figure 15), and the pH affect stability in a different way according pH, DH% and the enzyme utilized.

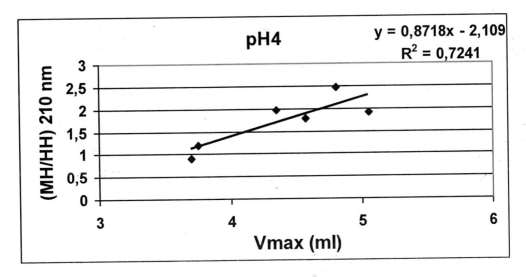

Figure 14. Relationship between maximal volume of liquid incorporated to the foam (Vmax.) of extracts at pH 4 and MH/HH (ratio of integrated areas of RP-HPLC chromatograms corresponding to medium hydrophobic and medium molecular weight components and high hydrophobic and high molecular weight components). Ac: acid protease. Al: alkaline protease.

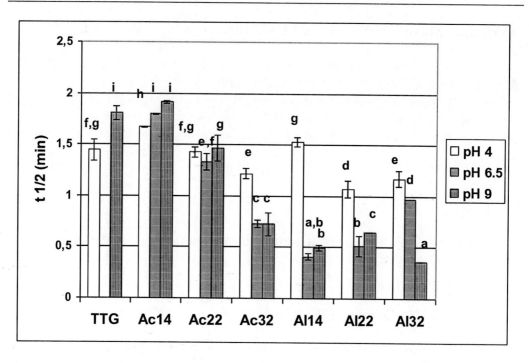

Figure 15. Time for the half-drainage of the liquid ($t_{1/2}$) corresponding to the different hydrolyzate extracts (protein concentration: 4 mg/ml) and TCAI % (14, 22 and 32). Ac: acid protease. Al: alkaline protease.

For Ac extracts, higher TCAI lower foam stability for the different pH extracts. This behavior is similar at that observed by Linares et al. (2000) from enzymic hydrolysates of wheat gluten in acid medium.

Many authors have observed that hydrolysis increase foam capacity but decrease stability (Popineau et al., 2002; Linares et a.l, 2000, Bombara et al.,, 1994, Mannheim and Cheryan, 1992). Protein partial cleavage and unfolding caused by hydrolysis, favour the rapid unfolding, rearrangements and the formation of a film in the water/air interface, but the film is not enough viscoelastic and thick to be stable (Schwencke, 1997).

a. Analysis of pH 4 Extracts

The better stability of Ac14 pH 4 extract with respect to TTG means that the components of the extract are more flexible and have better H/H balance than TTG. Also, the higher stability of Ac14 pH 9 foams imply that at this pH it is possible to extract components with an adequate H/H balance, capable to interact among them, because at pH 4, Ac14 extracts have components more hydrophobic and bigger than pH 9.

For Ac22, foam stability is similar for the different pH extracts but for Ac32 extracts, foams are more stable at pH 4 than pH 9 or 6.5. With acid extraction it is possible to obtain higher content of MH components (58.13 vs. 42.91 and 53.92) than at the other pH.

For Al extracts, foam stability was always better at pH 4 than at any other pH. Foams made with Al extracts were more unstable than with TTG extracts, except for Al 14 at pH 4.

At pH 4, a direct correlation between PCL and $t_{1/2}$ (r^2: 0.8504) was observed. At high PCL value, higher was the time of foam stability. At this pH, Ac 14 and Ac 22 have lower Ri than Al extracts, but have better stability. This could mean that pH 4 extracts are composed for a big peptide, which rearranged at the interface more slowly but then foams are more stable because they are formed by more structured components.

b. Analysis of pH 6.5 Extracts

For Ac extracts, foam stability decreases as DH increases, and foams made with hydrolysates from both enzymes have equal or lower $t_{1/2}$ than at another pH. However, Cheftel et al, (1989) have observed that foam stability is good when it is made with protein solutions at the pH near the isoelectric point (pI), like globin (pH 5-6), gluten (pH 6.5-7.5) and whey proteins (pH4 -5). This could be explained by the fact that extensive protein-protein interactions at pI, increase the thickness and strength of adsorbed protein at interface air/water, leading to strengthened films entrapping and stabilizing more air in the foam (Phillips et al., 1991; Kim y Kinsella, 1985). Also, surface tension has a minimum value near isoelectric point, owing to the minimum occupied area by a protein molecule. The highest surface viscosity and rigidity is observed near gluten pI, conducing to higher foam stability (Mita et al., 1977, 1978). However, these facts could be modified as a consequence of the effect made by the protein cleavage. In this regard, Popineau et a., (2002), observed that pH affects foaming capacity from ultrafiltrated hydrolysed wheat gluten having DH: 2.6% and medium molecular weight, 17 kDa. Permeates, rich in hydrophilic and neutral peptides, showed foaming capacity at pH 6.5 but not stabilization ability and at pH 4 foaming capacity was negligible. But the retentate contains peptides which are both hydrophobic and positively charged, which showed better foaming capacity and foam stability at pH 4 than 6.5.

In the case of Al extracts, foam stability improved with the increase of DH. It could be related to the enrichment with fractions not soluble at low TCAI.

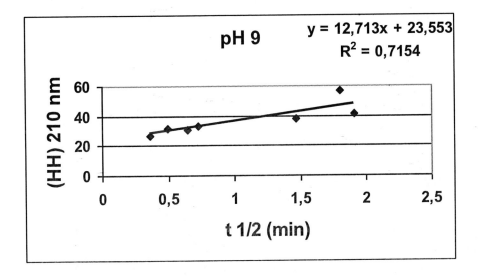

Figure 16. Relationship between time for the half-drainage of the liquid ($t_{1/2}$) corresponding to extracts at pH 9 and HH (integrated areas of RP-HPLC chromatograms corresponding to high hydrophobic and high molecular weight components).

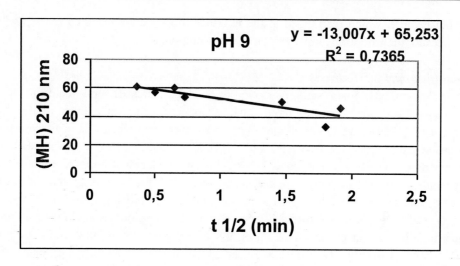

Figure 17. Relationship between time for the half-drainage of the liquid ($t_{1/2}$) corresponding to extracts at pH 9 and MH (integrated areas of RP-HPLC chromatograms corresponding to medium hydrophobic and medium molecular weight components). Ac: acid protease. Al: alkaline protease.

c. Analysis of pH 9 Extracts

For Ac extracts foam stability decreases as DH increases, but foams made with Al extracts, not only are less stable, but also the effect of DH is not clear, being foam made with Al 22 is more stable.

Regarding relationships among $t_{1/2}$ and the integrated areas of RP-HPLC chromatograms, foam stability was higher at higher HH content (Figure 16) and was lower when MH content increases (Figure 17). Then at pH 9, better capacity to form and stabilize foam is related to the presence of flexible HH components, being important for flexibility, size and H/H balance.

Analysis of the Initial rate of Liquid Drainage (V_0dren)

Once bubbling stopped, the liquid incorporated to foam start the drainage, depending on the foam stability, and a diminution of foam volume is observed along the time. The values of V_0dren are shown in Figure 18, where negative sign was not considered, and the higher value of V_0dren is related to a worse stability.

Foams made with TTG and Ac extracts have the same V_0dren at pH 4 and 9 and the drainage was independent of DH. At pH 6.5, V_0dren was higher than at any other pH and increases with DH.

Also, foams made with Al extracts have more drained liquid at pH 6.5. Near pI, enhanced hydrophobic interactions between two adsorbed protein layers caused by charges suppression make the lamellae between bubbles more thin and increasing V_0dren.

The behavior of Al extracts is different because V_0dren was higher for foams made with pH 6.5 and 9 extracts from Al 14 and decreases with the degree of hydrolysis. But at pH 4 Al 14 has the lower V_0dren.

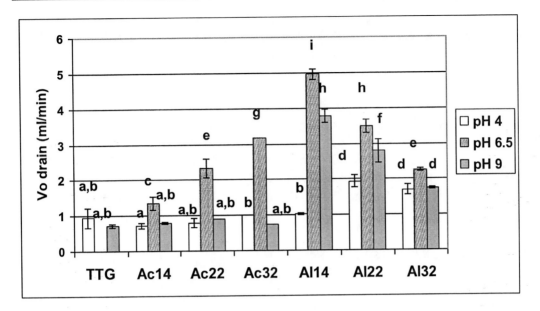

Figure 18. Initial rate of liquid drainage (V_0drain) corresponding to TTG (thermal treated gluten) and the different hydrolyzate extracts (protein concentration: 4 mg/ml); TCAI %: 14, 22 and 32. Ac: acid protease. Al: alkaline protease.

The enhanced stability related to the drainage of liquid at pH 9 for higher DH, could be due to interactions among peptides forming film mediated by SH/SS interchanges supported by alkaline pH (Roy et a.l, 1999). As it was observed in Figure 6, thiol groups content at pH 9 increases with DH. Damodaran (1994) observed a retardation of liquid drainage in the presence of higher concentrations of DTT (dithiotreitol), in foams made with bovine albumin, related to cleavage of increased number of disulfide bonds, and consequently, an increase in the hydrodynamic size of the protein, which might facilitate formation of a cohesive viscoelastic film. Also, German and Philips (1994) observed that disulfide bonds reduce the intrinsic flexibility of a native protein and SH/SS interchange is very important in manipulating functional properties. In the case of soy protein, disulphide bonds not only limit molecular flexibility but also restrict foaming. Molecular alterations induced by reducing inter-subunit disulfide bonds greatly improved film formation, foaming and foam stability.

Then, at pH 9, lack of structure cause by hydrolysis is compensated by the nature of extracts components related to their capacity of SS/SH interchange and the hydrodynamic size of peptides with higher thiol content.

No correlation was observed among V_0dren and integrated areas of RP-HPLC chromatograms.

Relationship among Indexes (RI, $t_{1/2}$, Vmax y V_0dren)

For foams made with extracts at pH 6.5 and 9, a direct correlation between Ri and t½ was observed (r: 0.7102 and 0.7534, respectively). This would imply that extracts containing fast arriving peptides which also interact themselves, would form the most stable foam. In the case of extracts at pH 4, no correlation was observed between these two indexes. This could be explained by the fact that at pH 4 the Ac extracts contain large molecules which form foams having low Ri but high t½. Furthermore at pH 6.5, an inverse correlation between V_0drain and Ri was observed (Figure 19), meaning that extracts which formed foams at higher rate, had less liquid drainage and had the highest t½. This was not observed at pH 9 since this tendency was only observed for Al and not for Ac extracts. It was also observed at pH 6.5 (Figure 20) a direct correlation between Vmax and $t_{1/2}$ meaning that those foams with more liquid incorporated during bubbling would be the most stable.

Regarding pH 4 extracts, it was observed that V_0dren directly correlated with Vmax (r: 0.8237), then being foams with more liquid incorporated which have the higher rate of liquid initial drainage. This is in accordance with the other results, since foams with higher Vmax were those made with Al extracts which are more unstable (less $t_{1/2}$) than those formed with Ac extracts.

Also at pH 6.5, the two indexes of foaming capacity are directly correlated, showing that foams with high Ri have high Vmax (Figure 21)

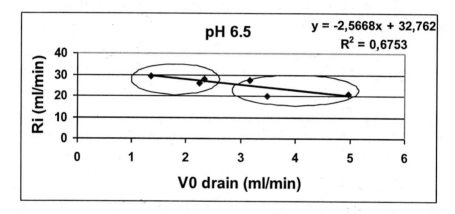

Figure 19. Relationship between initial rate of liquid drainage (V0drain) and the rate of incorporation of liquid to foam (Ri) corresponding to extracts at pH 6.5.

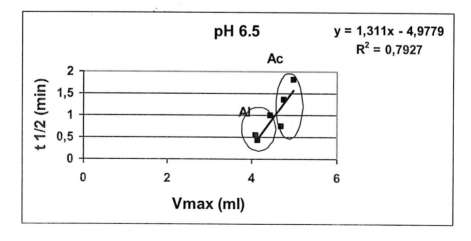

Figure 20. Relationship between time for the half-drainage of the liquid (t1/2) and the maximal volume of liquid incorporated to the foam (Vmax) corresponding to extracts at pH 6.5.

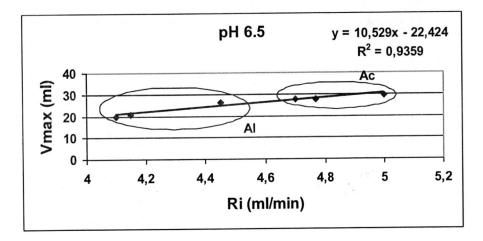

Figure 21. Relationship between rate of incorporation of liquid to foam (Ri) and maximal volume of liquid incorporated into the foam (Vmax) corresponding to extracts at pH 6.5.

On the other hand, the stability indexes (half life times and initial rate of drainage) correlated at pH 4 (r: 0.6517) and 6.5 (r: 0.815) but not at pH 9 (Figure 22). V_0dren inversely correlated with $t_{1/2}$, meaning that, foam with higher half life time, lesser liquid will drain.

In order to find other correlations among the different indexes, results corresponding for each enzyme, were analyzed separately. For Ac extracts a direct correlation between LH area (0-20 min) and V_0dren was observed (r: 0.6692), which would imply that the presence of peptides of low molecular weight and low hydrophobicity would contribute to the foam liquid drainage and to foam instability. Adler-Nissen and Olsen (1979) probed that small peptides and free amino acids impair stability of foams made with protein hydrolisates. Also, Berot *et a.*, (2001) observed that hydrophilic peptides from hydrolyzed wheat gluten reduced with cystein did not stabilize foams, although more hydrophobic peptides have better foam properties.

For the three pH Al extracts (Figure 23), a direct correlation between Ri and $t_{1/2}$, was observed (discarding TTG extracts which have low Ri but good stability, due to the presence of proteins of high MW) and also an inverse correlation between V_0drain and $t_{1/2}$ (r: 0.9013).

CONCLUSION

Foaming properties (foaming capacity and stability) depend on pH, DH and the enzyme used to produce the hydrolysates.

Both parameters of foaming capacity (Ri and Vmax) revealed that at pH 4, components of medium MW and hydrophobicity are related with better foaming capacity, indicated for a high rate of incorporation of liquid into the foam and higher volume of liquid. Acid extracts of hydrolysates have better properties than gluten without hydrolysis.

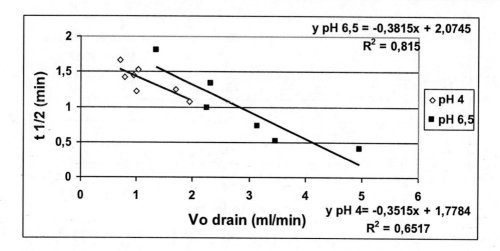

Figure 22. Relationship between initial rate of liquid drainage (V0drain) and time for the half-drainage of the liquid (t1/2) corresponding to extracts at pH 6.5 and 4.

At pH 9, both foaming capacity and stability are related to the presence of high hydrophobic peptides. Although at pH 4 a bigger molecule gives more stability, at pH 9 the smaller size molecules produced by hydrolysis, could be compensated by SS/SH interchanges, and by a bigger hydrodynamic size due to the peptides with free SH. Moreover, this could be reaffirmed by the fact that foam made with Al 14 at pH 4 has less V_0dren, but at pH 6.5 and 9 V_0dren decreases with DH, which is related with the enhanced content of thiol groups as TCAI increases in the Al extracts.

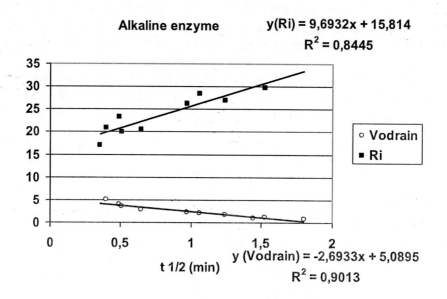

Figure 23. Relationship between the rate of incorporation of liquid to the foam (Ri) and the time for the half-drainage of the liquid ($t_{1/2}$) and the initial rate of liquid drainage (V_0drain) and time for the half-drainage of the liquid ($t_{1/2}$) corresponding to Alkaline protease hydrolysate extracts.

Some parameters correlated themselves, depending on the considered extracts. At pH near of pI (pH 6.5) a direct correlation between Ri and t½ and an inverse correlation between Ri and V₀drain exists. The extracts with higher Ri were the more stable (measured by the two stability parameters).

Vital gluten has good functional properties but the low solubility limits its use as a food ingredient. Mild hydrolysis gives the possibility of increasing its use by increasing solubility and functionality, but the final properties of hydrolysates depend on DH, enzyme and pH of functional property evaluation and the assay or parameter used. Both, charge and peptide composition depend on pH extraction and the hydrophilic/hydrophobic balance. Fractionating by pH gives peptide extracts with good foaming properties, like Ac extract at pH 9 or Al extract at pH 4 from hydrolysates of low DH (TCAI 14%).

ACKNOWLEDGMENT

Partially support by Proy. CAID + D 2005-005-25.

REFERENCES

AACC (1983) Approved Methods of the American Association of Cereal Chemists. 8th Edition.

Adler-Nissen, J. (1986) *Enzymic Hydrolysis of Food Proteins*. Elsevier Applied Science Publishers, London.

Adler-Nissen, J. and Olsen H.S. (1979) The influence of peptide chain length on taste and functional properties of enzymatically modified soy protein. *ACS Symposium Series*. 92, 125-146.

Anderson, W.L. and Wetlaufer, D.B. (1975) A new method for disulfide analysis of peptides. *Anal. Biochem.* 67, 493- 502.

Babiker, E.F.; Fujisawa, N.; Matsudomi, N.; and Kato, A. (1196) Improvement in the functional properties of gluten by protease digestion or acid hydrolysis followed by microbial transglutaminase treatment. *J Agric. Food Chem.* 44, 3746-3750.

Beveridge, T.; Toma, S.J.; and Nakai, S. (1974) Determination of SH- and SS-groups in some food proteins using Ellman's reagent. *J. Food Sci.* 39, 49-51.

Berot, S.; Popineau, Y.; Compoint, J.P.; Blassel, C.; Chaufer, B. (2001) Ultrafiltration to fractionate wheat polypeptides. *J. Chromatography B.* 753, 29-35.

Bombara, N.; Pilosof, A.M.R.; and Añón, M.C. (1994). Mathematical model for formation rate and collapse of foams from enzyme modified wheat flours. *J. Food Sci.* 59 (3), 626-628 y.681.

Cheftel, J.C., Cuq, J.L., Lorient, D. (1989). Propiedades funcionales de las proteínas. In: *Proteínas Alimentarias*. Ed. Acribia S.A. Zaragoza, España. Chap. 4, pp. 49-105.

Chobert, J.M.; Briand, L.; Gueguen, J.; Popineau, Y.; Larré, C.; and Haertlé T. (1996) Recent advances in enzymatic modifications of food proteins for improving their functional properties. *Nahrung.* 40 (4), 177-182.

Damodaran, S. (1994) Structure-Function Relationship of Food Proteins. In: *Protein Functionality in Food Systems.* Ed. Hettiarachchy, N.S. Ziegler G.R. IFT Basic Symposium Series 9, Chicago, Illinois. Chap. 1, pp. 1-37.

Drago, S.R. and González, R.J. Foaming properties of enzymatically hydrolysed wheat gluten. *Innovative Food Science & Emerging Technologies* 1: 269-273 (2001).

Drago, S.; González, R.J. and Añón, M.C. (2008) Application of surface response methodology to optimize hydrolysis of wheat gluten and characterization of selected hydrolysate fractions. In press. *J. Sci. Food Agricult.* 00-00.

German, J.B. and Phillips, L. (1994) Protein Interactions in Foams: Protein-Gas Phase Interactions. In *Protein Functionality in Food Systems.* Hettiarachchy, N.S. Ziegler G.R. (Ed). IFT Basic Symposium Series 9, chap 6, pp. 181-208.

Graham, D.E. and Philips, M.C. (1979a) Proteins at liquid interfaces: I-Kinetics of adsorption and surface denaturation. *J.Colloid Interface Sci.* 70, 403-414.

Graham, D.E. and Philips, M.C. (1979b) Proteins at liquid interfaces: II-Adsorption isotherms. *J. Colloid Interface Sci.* 70, 415-426.

Graham, D.E. and Philips, M.C. (1979c) Proteins at liquid interfaces: II I-Molecular structures of adsorbed films. *J. Colloid Interface Sci.* 70, 427-439.

Graveland, A.; Bosveld, P.; Lichtendonk, W.J.; Marseille, J.P.; Moonen, J.H.E.; and Scheepstra, A. (1985) A model for the molecular structure of the glutenins from wheat flour. *J Cereal Sci.* 3, 1-16.

Kella, N.K.D. and Kinsella, J.E. (1985) A method for the controlled cleavage of disulfide bonds in proteins in the absence of denaturants. *J Bioch. Biophysical Methods.* 11, 251-263.

Kim, S.H. and Kinsella, J.E. (1985) Surface activity of food proteins: relationships between surface pressure development, viscoelasticity of interfacial films and foam stability of bovine serum albumin. *J. Food Sci.* 50, 1526-1530

Kokini, J. and van Aken, G. (2006) Discussion session on food emulsions and foams. *Food Hydrocolloids.* 20, 438-445.

Kinsella, J.E. (1981) Relationships between structure and functional properties of food proteins. In: Food Proteins. Ed. Fox, P.F. y Condon J.*J.Applied Sci. Publishers.* London 3, 51-243.

Kinsella, J.E. and Phillips, L.J. (1989) Structure function relationships in food proteins: Film and foaming behavior. In: *Food Proteins: Structure and Functional Relationships,* J.E. Kinsella and W. Soucie (Ed). Am. Oil. Chem. Soc., Champaign, I.L. pp. 52-77.

Le Meste, M.; Colas, B.; Simatos, D.; Closs, B.; Courthaudon, J.L.; and Lorient, D. (1990) Contribution of protein flexibility to the foaming properties of casein. *J Food Sci.* 55 (5): 1445-1447.

Linares, E.; Larré, C.; Lemestre, M.; and Popineau, Y. (2000) Emulsifying and foaming properties of gluten hydrolysates with an increasing degree of hydrolysis: role of soluble and insoluble fractions. *Cereal Chem.* 77 (4): 414-420.

Linares E, Larré C, Popineau Y Freeze or spray-dried gluten hydrolysates. 1. Biochemical and emulsifying properties as a function of drying process. *J Food Engineering.* 48: 127-135 (2001).

Mac Ritchie, F. (1978) Proteins at Interfaces. *Adv. Prot. Chem.* 32, 283.

Mannheim, A. and Cheryan, M. (1992) Enzyme modified proteins from corn gluten meal: preparation and functional properties. *JAOCS* 69 (2), 1163-1169.

Mita, T.; Ishida, E.; and Matsumoto, I.I. (1978). Physicochemical studies on wheat protein foams. *J. Colloid Interface Sci.* 64 (1), 143-153.

Mita, T.; Nikai, K.; Hiraoka, T.; Matsuo, S.; and Matsumoto, H. (1977) Physicochemical Studies on Wheat Protein Foams. *J. Colloid Interface Sci.* 59 (1), 172-178.

Mutilangi, W.A.M.; Panyam, D.; and Kilara, A. (1996) Functional properties of hydrolysates from proteolysis of heat-denatured whey protein isolate. *J. Food Sci.* 61 (2), 270-303

Nakai, S. (1983) Structure-function relationships of food proteins with an emphasis on the importance of protein hydrophobicity. *J Agric. Food Chem.* 31, 676-683.

Nielsen, P.M.; Petersen, D.; Dambmann, C. (2001) Improved method for determining food protein degree of hydrolysis. *J Food Sci.* 66, 642-646

Phillips, L.G.; Yang, S.T.; and Kinsella J.E. (1991) Neutral salt effects on stability of whey protein isolated foams. *J Food Sci.* 56 (2), 588-589.

Popineau, Y.; Huchet, B.; Larré, C.; and Bérot, S. (2002) Foaming and emulsifying properties of fractions of gluten peptides obtained by limited enzymic hydrolysis and ultrafiltration. *J. Cereal Sci.* 35, 327-335.

Roy, S.; Weller, C.L.; Gennadios, A.; Zeece, M.G.; and Testin, R.F. (1999) Physical and molecular properties of wheat gluten films cast from heated film-forming solutions. *J. Food Sci.* 64 (1), 57-60.

Schwencke, D.D. (1997) Enzyme and chemical modification of proteins. In: Food proteins and their applications. Ed. Damodaran S y Paraf A. Marcel Dekker, New York, 393-423.

Tsaliki, E.; Pegiadou, S.; and Doxastakis, G. (2004) Evaluation of the emulsifying properties of cottonseed protein isolates. *Food Hydrocolloids.* 18, 631-637.

van der Ven, C.; Gruppen, H.; de Bont, D.B.A.; and Voragen, A.G.J. (2001) Reversed phase and size exclusion chromatography of milk protein hydrolysates: Relation between elution from reversed phase column and apparent molecular weight distribution. *Inter. Dairy J.* 11, 83-92.

Wagner, J.R.; Sorgentini, D.A.; and Añón, M.C. (1996) Thermal and electrophoretic behavior, hydrophobicity and some functional properties of acid-treated soy isolates. *J Agric. Food Chem.* 44, 1881.

Chapter 9

GLUTEN-FREE DIET IN CHILDREN AND ADOLESCENTS WITH CELIAC DISEASE[*]

Gian Vincenzo Zuccotti, Dario Dilillo, Fabio Meneghin and Cecilia Mantegazza
University of Milan, Luigi Sacco Hospital, Italy

ABSTRACT

Celiac disease (CD) is defined as a permanent sensitivity to the gluten in wheat and related proteins found in barley and rye, which occurs in genetically susceptible individuals and affects 0.5-1% of the general population worldwide. Currently, the only available treatment is lifelong adherence to a gluten-free diet (GFD). There is evidence that untreated CD is associated with a significant increase in morbidity and mortality. It has been demonstrated that even small amounts of ingested gluten can lead to mucosal changes upon intestinal biopsy. Previously, products containing less than 200 ppm (<200 mg/kg) were regarded as gluten-free. Currently, less than 20 ppm (<20 mg/kg) is being considered in the proposed Codex Alimentarius Guidelines to define gluten-free. In the USA, the national food authority has recently redefined their definition of "gluten-free" with a threshold of no gluten. The use of oats is not widely recommended because of concerns about potential contamination during the harvesting and milling process, so unless the purity of oats can be guaranteed, its safety remains questionable.

A GFD has both lifestyle and financial implications for patients; thus, it can adversely impact their quality of life, such as difficulties eating out, a negative impact on career and family life, anxieties about social difficulties, and feeling different. Though the compliance to a GFD started in childhood is very high, the percentage of adolescents with CD who strictly follow a GFD varies from 43% to 81%; a greater adherence is found within patients with typical symptoms. Moreover, the alimentary habits of healthy adolescents exhibit nutritional imbalances with a high consumption of lipids and proteins and a low consumption of carbohydrates, calcium, fiber and iron; several studies show

---

[*] Chapter 9 – A version of this chapter was also published in *Celiac Disease: Etiology, Diagnosis, and Treatment*, edited by Matthew A. Edwards, published by Nova Science Publishers, Inc. It was submitted for appropriate modifications in an effort to encourage wider dissemination of research.

that adherence to a strict GFD worsens the nutritional imbalances of an adolescent with CD.

Gluten-free products are often low in B and D vitamins, calcium, iron, zinc, magnesium, folate, thiamine, riboflavin, and niacin. Very few commercially available gluten-free products are enriched. Vitamin and mineral supplementation can be useful adjunct therapy for a GFD. Patients inadequately treated have low bone mineral density, imbalanced macronutrients, low fiber intake, and micronutrient deficiencies.

Moreover, recent research has found that adults on a strict GFD for years have high total plasma homocysteine levels; this finding is associated with a high prevalence of being overweight and represents an increased risk for metabolic and cardiovascular disease.

Celiac disease (CD) is an immune-mediate enteropathy caused by a permanent sensitivity to gluten in genetically susceptible individuals, in whom the proximal small bowel mucosa is damaged as a result of dietary exposure to gluten. The prevalence of CD varies in different regions of the world, but has been estimated to have a worldwide prevalence of at least 1%. These findings suggest that CD may be one of the most common chronic disorders [1]. Large-scale screening studies in the adult population suggest that the diagnosed celiac population may represent a small fraction of a bigger problem; in fact, the asymptomatic form of CD, with positive serological and histological findings, may be five to seven times more common than the symptomatic disease [2].

Celiac disease is defined as a permanent sensitivity to the gluten in wheat and related proteins found in rye and barley. The activity of gluten resides in the gliadin fraction, which contains certain repetitive amino acid sequences that lead to sensitization of lamina propria lymphocytes. This immunological response involves the activation of CD4+ gluten-sensitive T cells, and the inflammatory response results in villous atrophy, crypt hyperplasia, and damage to the surface epithelium of the small bowel [3,4].

The only treatment currently available for CD is a strict and lifelong adherence to a gluten free diet (GFD). Prolonged adherence to a GFD may reduce the morbidity and mortality associated with diagnosed but untreated CD, such as infertility, recurrent abortions, low birth weight, intrauterine growth retardation, early menopause, skeletal disorders, short stature, chronic fatigue, malignancy, and many others [5,6]. In the United States, still, it is estimated that symptoms are present for a mean of 11 years before diagnosis [7]. A Swedish study found an elevation in mortality from all causes in a group of 828 celiac patients and found risks elevated in multiple areas, such as non-Hodgkin's lymphoma, small intestinal cancer, rheumatoid arthritis, connective tissue disease, and diabetes [8]. An English study reported a 30% overall increased risk of any malignancy in a large group of celiac patients [9].

The complications of CD, such as malignancy, typically occur many years after diagnosis and usually are observed in adults. The primary etiology for increased mortality is the association with gastrointestinal malignancies, in particular intestinal lymphoma, which is reported in 10-15% of adult patients with CD who have been noncompliant with a GFD. Most studies report an increased risk of non-Hodgkin lymphoma, but they often do not distinguish between the classic celiac associated lymphoma (EALT; enteropathy associated T-cell lymphoma) and other subtypes. However, even with these increased risks, this important complication is very rare. There is also an increased risk of adenocarcinoma of the small intestine and of other carcinomas in the gastrointestinal tract [10,11].

There are two major categories of bone diseases seen in celiac disease: osteoporosis, which is quite common, and osteomalacia, which is relatively infrequent and characterized by bone pain, muscle weakness, pseudofractures, high serum levels of alkaline phosphatase, and a low vitamin D concentration. The main cause of this disease is malabsorption; contributing factors may include inadequate calcium and vitamin D intake, corticosteroid use to treat coexisting morbid autoimmune disorders or refractory sprue, or general risk factors such as aging, alcohol, smoking, inactivity, and low body weight [12,13].

Osteoporosis has taken the lead as the chief cause of morbidity in patients with untreated CD. About 40% of patients with newly diagnosed CD have osteopenia and 26% have osteoporosis, but strict adherence to a GFD will lead to a significant improvement in bone density [14,15]. Fertility problems, sexual dysfunction, and obstetrical complications are more frequently observed in patients with CD. These reproductive disorders may be a consequence of the endocrine derangements caused by selective nutrient deficiencies. Women diagnosed with CD more frequently experience recurrent spontaneous miscarriages, delayed menarche, early menopause, amenorrhea, and vaginal discharge. Fertility problems are also common in both men and women. In about 25% of celiac patients, hyperprolactinemia is diagnosed, which may be one of the causes of impotence and the loss of libido [16-19]. For these reasons, a diagnosis and treatment with a GFD is desirable as early as possible.

There is a lot of evidence that the lack of dietary compliance is most often due to inadequate education about the GFD, misinformation, and also the diet's complexity. Even the most motivated and highly educated patients can have difficulty adjusting to a GFD. For this reason, it is very important for all CD patients to follow intense expert dietary counseling prior to starting a GFD. Other than the education of patients, educating family members about CD and the GFD can improve compliance.

DIETARY INSTRUCTION

The dietitian and the patient's physician who counsel patients with CD should have expertise in this disease and the GFD. Dietary education should begin with reinforcement of the importance of maintaining a strict gluten-free lifestyle, including a review of the pathophysiology in prompt terms and a nonthreatening discussion of the long-term consequences of not maintaining a GFD. Patients who do not experience symptoms after the ingestion of gluten should be reminded that damage may still occur despite the lack of outward symptoms [20].

Topics to be covered in dietary education are:

- *At initial assessment: weight, weight history, diet history, 24-h recall, supplements, and nutrient deficiencies.* The initial diet consultation should include an assessment of the patient's present state of health, eating habits, lifestyle, and GFD instruction. If a patient has lost weight or is underweight, treatment with a GFD should result in spontaneous weight gain as the associated malabsorption is corrected. A diet history should include a review of the patient's typical diet, where meals are eaten, and who purchases and prepares food. Clinicians need to be sensitive to the emotional and psychological effects that diagnosis and treatment can have on an individual [21]. It is important to assess the patient's acceptance of the diagnosis and willingness to change their eating habits. Everyone could have a different reaction to diagnosis and it is common for such patients to react with anxiety, anger, or depression. It is important to explore the patient's support system and involve any others near to the patients in the education [22]. Clinicians have to encourage patients to go on living as normally as possible with the GFD. Patients who present with nutrient deficiencies may require temporary or long-term nutrient supplementation (these deficiencies often correct themselves spontaneously with the GFD). The most common deficiencies are of iron, calcium, folate, and vitamins. Calcium and vitamin D supplementation may be needed long-term if bone disease is present. In the case of nutrient deficiencies, a close follow-up is important for assessing the correction [23].

- *At counseling*: *importance of strict compliance, sources of gluten (food and other products), alternative gluten-free food, where to purchase alternative food, support groups, label reading, and eating away from home.* A list of gluten-containing grains and gluten-free grains can be seen in Table 1 below. All patients should be instructed to avoid toxic grains and all products made from these grains, such as pasta, cereals, gravies, sauces, breads, pastries, pie crusts, cakes, cookies, crackers, soup, and any food with gluten-containing additives [24]. Most plain meats, vegetables, fruits, and rice are inherently gluten-free as long as they are prepared without contamination with gluten-containing ingredients. In the last few years, alimentary industries have put forth a lot of effort to improve the availability and quality of foods using alternative flours. Actually, celiac patients are able to purchase a gluten-free substitute for almost any gluten-containing food, including pizza, pasta, and even beer [25]. Because anything packaged may be a potential source of gluten, patients need to check the commercial foods carefully, reading the label well, purchase a shopper's guide, and contact the manufacturer directly. An additional difficulty for the patients is that many derivatives of gluten-containing grains are used as additives, preservatives, and stabilizers in packaged and processed foods [26]. Ingredients that potentially contain gluten are malt or malt flavoring, modified food starch, flavorings, dextrins, and seasonings. Patients should be instructed to avoid eating a food with questionable ingredients until its safety can be verified. Another troublesome aspect of the GFD is eating away from home, including not only the restaurant, but also travelling by car or air, visiting someone else's home, or going to school. Local support groups are often available that can be contacted about restaurants that accommodate gluten-free food. A potential source of gluten in restaurant meals, and even at home, is cross-contamination. Sources of contamination at home may include condiment containers (jam, margarine, utensils), countertops, cutting boards, and the toaster. Outside gluten-free foods may be contaminated in fields where grains are grown and harvested, in mills where grain is processed into flour, or on food-processing lines where one line produces gluten-free products and another produces gluten-containing ones. The best way to ensure compliance from children is to educate and empower them right from the beginning. They should not be denied going to sleepovers, birthday parties, or other events. Non-food sources of gluten may be toothpaste, mouthwash, lipstick, and postage stamps, but also prescription medications and vitamin and mineral supplements that may contain gluten in their inert ingredients, excipients, coatings, or capsules. Potential sources of gluten in medications include starch, dextrins, and other ingredients [27].

Table 1. Modified from See J. Gluten-free diet: the medical and nutrition management of celiac disease. *Nutrition in Clinical Practice* **2006, 21, 1-15**

Classification of grains according to gluten content	Gluten-free grains
Barley	Amaranth
Bran	Buckwheat (kasha)
Bulgur	Corn
Couscous	Flax
Durum	Millet
Einkorn wheat	Montina flour
Emmer wheat	Oats
Faro	Popcorn
Farina	Potato flour
Graham flour	Quinoa
Kamut	Rice, brown
Orzo	Rice, white
Rye	Rice, wild
Spelt	Sorghum
Triticale	Soybeans
Wheat	Teff
	Tapioca

- *At follow-up*: evaluate the weight, compliance, comprehension, dietary adequacy, and variety and troubleshooting (e.g., the intentional or unintentional ingestion of gluten). Follow-up visits with the dietitian are essential to assess knowledge, competence, and compliance. The first visit may be scheduled within one to three months. After the start of a GFD, a weight change may be an indicator of compliance or response to treatment, but it must be interpreted with caution. Weight stabilization or gain is a good sign that the patient is responding to the diet. If a patient continues to lose weight, or if an underweight patient fails to gain weight, it may be due to noncompliance, refractory CD, or inadequate caloric intake. An efficient way to assess comprehension of the GFD, dietary adequacy, and compliance is the food record in which the patient had to keep a diary of brand names, method of preparation, and where the food was eaten. Using this food record, a dietitian can observe possible intentional intake and evaluate their compliance to the diet. The Celiac Disease Guideline Committee recommends the measurement of anti-transglutaminase tissue antibodies (tTG) after 6 months of treatment with the GFD to observe a decrease in the antibody titer as an indirect indicator of dietary adherence and recovery. The measurement of tTG is also recommended for individuals with persistent or recurrent symptoms at any time after starting a GFD, as a rise in antibody levels suggests dietary nonadherence [28]. Regarding dietary adequacy, it is common in CD patients to find lactose intolerance, often secondary to lactase deficiency as a result of injury to the gastrointestinal tract. This lactase deficiency often disappears when the intestines heal, but it can be persistent in some patients; in this situation, low-lactose dairy products, such as aged cheese, yogurt with active cultures, and lactose-treated milk, may be well tolerated. If patients are unable to consume adequate amounts of calcium and vitamin D, supplements should be prescribed. Studies show that patients with CD often have a low intake of B vitamins, fiber, and iron, because all of these are found in significant amounts in grain products, but most of them are restricted on a GFD. Furthermore, some gluten-free grain products are lower in these nutrients. It is important to encourage patients to ingest adequate amounts of these nutrients because they are important to prevent heart disease, diabetes, cancer, and obesity [29-31].

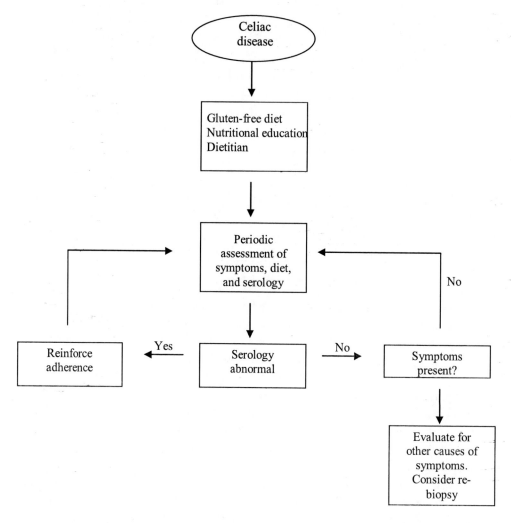

Figure 1. Flowchart of follow-up strategies. Modified from Guideline for the diagnosis and treatment of celiac disease in children: recommendations of the North American Society for Pediatric Gastroenterology, Hepatology and Nutrition, (2005), *Journal of Pediatric Gastroenterology and Nutrition, 40,* 1-19.

Most patients who have been properly instructed will respond to the GFD and symptoms will start to improve almost immediately. Management of nonresponsive CD should begin with a very detailed assessment of dietary intake to detect any sources of gluten [32]. If no gluten ingestion is detected by this assessment or by measuring serological markers, a workup should ensure the identification of coexisting causes of symptoms, which could include pancreatic insufficiency, lactose or other food intolerances, irritable bowel syndrome, bacterial overgrowth, collagenous colitis, lymphoma, or parasites. Antigliadin antibodies are still useful to detect gluten contamination in treated patients because they are more sensitive than anti-tissue transglutaminase antibodies that remain or become positive in the setting of substantial gluten intake of approximately 1 g or more daily. *Refractory celiac disease* can be defined as continued diarrhea, malabsorption, and villous atrophy associated with progressive malnutrition despite a rigorous GFD for over 6 months [33]. Patients with this condition should continue to follow a strict GFD, but it will require intensive nutritional support and amelioration of the secondary effects of the severe malnutrition that can persist or develop. Some patients will respond to immunosuppression with corticosteroids or azathioprine. Severe malnutrition as indicated by persistent weight loss and nutrition deficiency is a common sequelae of nonresponsive disease and is an independent predictor of mortality in patients with refractory sprue. Parenteral nutrition may have a role in the management of these patients.

OATS

Cereal grains are taxonomically classified into the Poaceae family, within which there are several subfamilies; one of these, the Festucoidae family, contains wheat, barley, oats, and rye. The proteins in these cereals that are responsible for their toxicity are known as prolamines, which are called gliadin in wheat, hordeins in barley, secalins in rye, and avenins in oats [34].

Several studies have shown that tissue transglutaminase (tTG), an enzyme responsible for the cross-linking of proteins, is responsible for the deamination of glutamine and proline residues in prolamines, and this is the trigger for the inflammatory process and enterocyte damage [35]. Recently, it has been hypothesized that patients with CD may be able to tolerate oats because of their lower prolamine content compared to other grains of the same family. In fact, the prolamine, gliadin, in wheat constitutes 40% of cereal, and the percentages are similar in rye and barley; however, in oats the prolamine avenin constitutes only 5-15% of the cereal [36]. Moreover, avenins contain fewer proline residues than the other cereals and, thus, offer fewer deamination sites for the tissue transglutaminase enzyme [37].

The inclusion of oats in the GFD would be beneficial because they provide a good source of fiber, have a higher satiety value than other gluten-free cereals, and increase patient choice. Patients with CD adhering to an oat-containing GFD have reported an increased diversity of diet with a decreased cost, increased availability of foods, and more palatability. Despite these preliminary remarks, the inclusion of oats in the GFD remains highly controversial.

So far, most of the studies have been focused on adult CD patients and have not shown any toxicity to oats in moderate or large amounts; furthermore, *in vitro* studies also indicate a lack of oat toxicity in CD patients. Although oats are generally considered safe for adults, the recommendations for children with CD have been to not include oats in the diet because of a lack of scientific evidence concerning safety.

Firstly, there is concern about the possible contamination of commercial oat products by other gluten-containing cereals during the harvesting and milling process. Secondly, it has been noted that a subgroup of celiac patients experience more abdominal symptoms while consuming an oat-containing GFD. Thirdly, it was recently demonstrated by Arentz-Hansen and her group [37] that, even if oats seem to be well tolerated by many CD patients, there are a small subset of celiac patients who cannot tolerate oats and have an important intestinal T-cell response. Until the prevalence of this oat intolerance in CD patients is established, clinical follow-ups of CD patients eating oats is advisable. Clinicians should be aware that oat intolerance may be a reason for villous atrophy and inflammation in patients with celiac disease who are eating oats but otherwise adhering to a strict GFD. It might be possible in the future to test patients with CD for oat intolerance by monitoring the T cell responses to avenin epitopes *in vitro*. Lastly, a fourth concern is that there are only a few studies regarding the use of oats in CD children.

The first randomized, double-blind, multicenter study of the suitability of oats for children [38] included 116 children with newly diagnosed CD. The results, possibly influenced by a high withdrawal frequency, indicate that the presence of moderate amounts of oat in a GFD does not prevent clinical or small bowel mucosal healing or humoral immunological down-regulation. This finding is in accordance with previous open studies in celiac children and adults. All of the studies performed to establish the safety of oats presented short-term results for diet test periods of up to 24 months. The absence of signs of toxicity to oats does not exclude the possibility of long-term complications.

A more recent study [39] found for the first time in CD children that the consumption of considerable amounts of oats is well tolerated and does not result in small bowel mucosal damage or a humoral response, even when taken for many years. In CD, the most sensitive indicator in a small bowel mucosal response to gluten has been an increase in intraepithelial lymphocytes (IEL), and a decrease in IEL number is the last value observed after mucosa normalization [40]. Even more reassuringly, in this study, after long-term consumption of an oat-containing GFD, all celiac children were found to have normal IEL. According to the current literature, some celiac patients experience gastrointestinal symptoms more often during an oat-containing, as opposed to a traditional, GFD [41]. These symptoms have been mild and the appearance of abdominal distension or flatulence has been explained by an increased intake of fiber from oat products [42]. The authors of this study concluded by saying that uncontaminated oats can be safely included in the GFD of a majority of celiac children. Celiac societies and industry should make an effort to produce oat products free of wheat contamination.

As avenin is less immunogenic than gluten, it might take a considerably longer time period to trigger a relapse. Maybe it would be prudent to suggest that patients wanting to include oats in their diet undergo annual biopsy screenings to ensure that subclinical changes are not occurring in the small bowel mucosa and potentially conferring an increased risk of developing complications associated with CD [43].

COMPLIANCE TO GFD AND QUALITY OF LIFE

Several studies have attempted to determine the effects of a GFD by examining food consumption and limitations on the quality of life for CD patients. The results of these studies show that a GFD impacts various lifestyle aspects of the quality of life for these individuals and that the dietary compliance frequently is not strict, especially in adolescents where the correct diet is often neglected. The percentage of adolescents who follow a GFD varies from 43% to 81%, according to several studies [44-47]. A recent study [48] showed that 49% of the respondents to a questionnaire, which was mailed to more than 400 members of a Celiac Disease Support Group, rated their health before starting a GFD as good, though 32% rated it as poor. In rating their current health, 71% thought it was excellent, with 60% saying they were as healthy as anyone else. Most of the respondents (81%) thought that CD has no effect on their ability to work. A lot of patients thought that neither their physical health (87%) nor their emotional well-being (90%) had any effect on their social activities. In contrast, they reported that maintaining a GFD had a negative impact on their quality of life; 82% reported a negative impact on travel, 67% on family life, 86% on dining out, and 41% on their careers. These results varied with gender because more women reported negative impacts on dining out (65%), travel (64%), and family life (49%) than men (20%, 18%, and 18%, respectively).

When queried about who instructed them about the GFD, patients answered that 71% obtained information from books, the internet, support groups, family, or friends and only 17% received information from a doctor or dietitian (13%).

In another study [49], which aimed to evaluate the factors involved in the impairment of health-related quality of life (HRQOL) in patients with CD, the findings show that patients in the GFD group reported a significantly better HRQOL than patients in the pre-treatment group for the overall score, gastrointestinal symptoms, and in the physical, social, and emotional dimensions (see Figure 2). Patients in the pre-treatment normal diet group, scored best on the gastrointestinal quality of life index (GIQLI) treatment effects domain, whereas the domains with the worst scores were physical and emotional dysfunction and gastrointestinal symptoms. In the GFD patient group, the best scored domain was also the treatment domain, and the worst scored domain was emotional dysfunction, which suggested that, although improved by treatment, the most sustained impairment was the emotional dimension. In confirmation of these results, in the GFD group, the problems reported most frequently were anxiety and depression (33.9%), less considerable was pain/discomfort (27.7%) and other aspects; in the normal diet group, the major problems reported were pain/discomfort (62%), anxiety/depression (54.5%), and, least frequently, the self-care dimensions (10.2%).

Despite 90% of celiac patients perceiving themselves as healthy and that their CD had little effect on their quality of life, those perceptions contrast sharply with the negative responses to specific quality of life questions, including dining out, travel, family, or career. The chronic nature of this disease, with the limitation of the GFD, substantial number of physician visits, the risk of associated disease, and potential complications means that CD can have a considerable negative impact on the HRQOL. In particular, the restrictive nature of the diet can be a limiting factor affecting decisions regarding social activities [50]. The most frequent sources of forbidden gluten were candies (53%), chocolates or crisps (47%), and fast food (31%); the gluten-containing food products were mostly consumed at special occasions

(60%) or at home (49%). A high percentage of patients (65%) reported symptoms after gluten consumption, especially abdominal pain (83%), diarrhea (73%), and lassitude (52%). It is important to understand that, although dietary non-compliance might be easier socially, there is an increased risk of acquiring diseases such as lymphoma, osteoporosis, and anemia. For these reasons, it is important to adhere to a strict lifelong GFD. The findings of these studies show that the assessment of the HRQOL questionnaire in CD patients is relevant because it improves physicians' knowledge of the implications of the disease in patients' daily life and also helps patients to recognize the general impact of the disease.

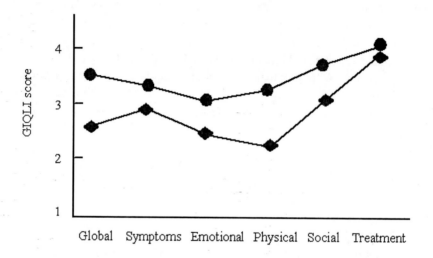

Figure 2. The gastrointestinal quality of life index (GIQLI) for celiac disease patients on a gluten-free diet (circles) and those pre-treated on a normal diet (diamonds). Modified from Casellas F. Factors that impact health-related quality of life in adults with celiac disease: a multicentre study. *World J Gastroenterol*. 2008.14(1):46-52.

TRESHOLD OF GLUTEN

A lifelong GFD is the only effective treatment of patients with CD. Estimation of the maximum tolerated amount of gluten for susceptible individuals would support the effective management of their disease. The GFD is defined in most countries in accordance with the Codex Alimentarius commission of the food and agricultural organization of the United Nations (FAO) and the World Health Organization (WHO). The Codex alimentarius allows the inclusion of up to 0.3% protein from gluten-containing grains in food labeled gluten-free [51]. Since 1995 in Australia, no detectable gluten (NDG) was introduced for a GFD with a lower limit of detection of 0.001% gluten [52].

The disappearance of symptoms should follow good compliance with a GFD, but this occasionally does not happen [53]. Although this could be attributed to conditions unrelated to CD, the trace amounts of gluten present in a Codex-GFD may also be responsible [54,55]. The potential toxicity of trace amounts of gluten is still unclear, and a wide individual variability exists. The question of whether all celiacs should follow an NDG-GFD or whether the trace amounts of gluten found in the Codex-GFD can be permitted in some or all patients remains controversial [56]. Certainly, there is now no doubt that such products can provoke symptoms in a subgroup of sensitive patients. On the other hand, no correlation between mucosal damage and the Codex-GFD has been found, nor a change in mucosal appearance after switching to a NDG-GFD.

In a GFD, the maximum amount of gluten should be between 10 and 100 mg daily. Different studies have tried to solve this problem: Ciclira et al [57] concluded that 10 mg produced no change, 100 mg a very slight measurable change, 500 mg a moderate change, and 1 g extensive damage to the morphology of the small intestine; Ejderham et al [58] showed that a daily intake of 4-14 mg gliadin did not affect the morphology of the small bowel mucosa in CD patients receiving a long-term treatment with a GFD; recent Finnish studies [59] indicate that an intake of 20-36 mg gluten daily has no detectable effect on mucosal histology; and Catassi et al [60] demonstrated that 50 mg gluten daily is the minimum dose required to produce measurable damage to the small intestinal mucosa in CD patients. This last study shows that a threshold of 20 ppm keeps the intake of gluten well below the amount of 50 mg/d, which allows a safety margin for the variable gluten sensitivity and dietary habits of patients. This prudent value has only been adopted in North America and southern European countries.

NUTRITIONAL IMBALANCE IN CD AND THE GFD

Several studies have shown the important nutritional imbalances in untreated or newly diagnosed CD patients. An Italian study [46] reported that 72% of adolescents with CD adhering strictly to the diet were overweight and consumed an unbalanced diet rich in fats and proteins, poor in carbohydrates, and deficient in calcium, iron, and fiber. Another study [61] performed in the Netherlands found that celiac patients aged 12-25 years old had an intake of fiber and iron significantly lower than the recommended dietary allowance (RDA) and an energy percentage of saturated fat significantly higher than the Dutch RDA (DRDA). In addition, the intake of calcium by patients younger than 19 years old did not reach the American RDA. One of the causes of these incorrect dietary habits is that celiac patients reduce their consumption of some foods containing carbohydrates, such as pizza, bread, or pasta, and do not counterbalance with an increase in alimentary items of the same group, such as rice, corn, and potatoes.

Because the proximal small bowel is the principal site of inflammation, but is also the site of *iron* absorption, there is an important iron malabsorption associated with a refractory iron deficiency anemia; the frequency in celiac disease varies from 12% to 69% [62]. The frequency of occult gastrointestinal bleeding is possibly a contributing factor in iron deficiency anemia [63]. Correcting an iron deficiency is achieved by following the GFD and taking ferrous iron. The most commonly used regimen for iron supplementation is ferrous sulphate at a dose of 200 mg three times daily. Iron depletion may take as long as 12 months to correct, although anemia is usually reversed by 6 months. Persistent anemia may be a sign of inadequately treated CD, but anemia sometimes persists in spite of excellent disease control and, in these cases, the physician should try to identify the cause. Potential causes include blood loss, non-compliance with iron supplements due to side effects, and ineffective absorption due to the concomitant use of medications that reduce stomach acid or otherwise interact with iron absorption.

Vitamin B_{12} deficiency is thought to be uncommon in CD because of the fact that the ileum, where this vitamin is absorbed, is relatively spared [64]. It is generally accepted that vitamin B_{12} malabsorption is restricted to patients with advanced disease involving the ileum, and that vitamin deficiency alone is seldom severe enough to cause anemia or neuropathy. Moreover, deficiencies in specific transport proteins may also contribute to a low vitamin level. In some cases, the apparent vitamin B_{12} deficiency may be secondary to folate depletion and vitamin levels may correct themselves after folate replacement, but recent studies show that the subnormal vitamin B_{12} levels were not always secondary to folate deficiency and were found alone. For these reasons, it is advisable to know the vitamin B_{12} status before hematinics are replaced, especially folate replacement. The incidence of vitamin B_{12} deficiency ranges from 11% to 41% [65]. Vitamin B_{12} deficiency will resolve on a GFD alone without parenteral supplementation, and such supplementation may be required only in those symptomatic for vitamin deficiency. In celiac patients presenting with classic malabsorption, deficiencies in fat-soluble vitamins (A, D, E, K) are commonly encountered. Clinical manifestations of this kind of deficiencies are night blindness, conjunctival dryness, and keratomalacia for vitamin A; osteomalacia (muscle weakness and musculoskeletal pain) for vitamin D; hemolytic anemia, neuropathy, and gait disturbance for vitamin E; and a prolonged prothrombin time and low levels of II, VII, IX and X factors predisposing to hemorrhage for Vitamin K [66-68].

Often, in additional nutritional screenings of CD patients copper, zinc, selenium, and folate deficiencies are found. Folic acid is absorbed in the jejunum, the proximal segments of which can be inflamed and damaged in active CD and is often present in red blood cell macrocytosis, although the presence of iron deficiency anemia may result in a mixed picture on a blood smear. For this reason, folate supplementation is recommended during a GFD [69,70]. Recent studies have spotlighted a low serum zinc concentration in patients with active CD. Zinc is an essential trace element that is required for DNA synthesis, cell division, protein synthesis, and immune response [71]. Approximately 300 enzymes are dependent on zinc for their activities. A mild deficiency state, often detected in newly diagnosed celiac patients, causes impaired immune function, hypogonadism, oligospermia, alopecia, poor wound healing, skin changes (pustular and vesiculobullous lesions), and a failure to thrive. Primary zinc deficiency has been associated with the impaired synthesis and secretion of follicle-stimulating hormone (FSH) and luteinizing hormone (LH), abnormal ovarian development, and obstetrical disorders such as spontaneous abortion, congenital malformations, stillbirth, pre-eclampsia, and intra-uterine growth retardation [72,73]. Even selenium deficiency is often diagnosed in celiac patients, and this element is important in the reproduction process; selenium is involved in spermatogenesis and a deficiency results in subfertility [74]. The exact prevalence of copper deficiency in celiac patients is unclear. A low serum copper level could be involved in hematological abnormalities, such as anemia, neutropenia, macrocytosis, and elevated serum ferritin and erythropoietin; these changes normalize with a copper repletion [75,76].

In contrast with well known disorders of calcium and phosphorous metabolism, the nutritional magnesium status in CD has been studied to a lesser extent [77]. The key role played by magnesium in the metabolism of proteins, nucleic acids, glucose, fat, and transmembrane transportation, and the fact that some symptoms of subclinical CD, such as tetany and tremors, can be associated with tissue magnesium deficiency, indicate that it is necessary to evaluate the Mg balance in subclinical CD. Magnesium absorption occurs mainly in the ileocolonic segment of the intestine and is greater when the supply of Mg is lower. In clinical CD, Mg deficiency is probably caused by accelerated passage in the digestive tract, steatorrhea, and lower food intake because of a lack of appetite. Important sources of dietary magnesium are cereal products, cocoa, sweets based on nuts, sunflower seeds, milk, and potatoes. In the GFD, the lower supply of magnesium is usually due to the lower Mg content in gluten-free cereal products compared to conventional ones [78].

Recent studies [32,79,80] show that only a minority of newly diagnosed celiac patients are underweight, with a high proportion having a body mass index (BMI) of 25 or over (defined as overweight or obese according to WHO criteria). One of these studies [81] found at diagnosis a mean BMI of 24.6, with a range of 16.3-43.5, for the celiac patients; about 5% were underweight, 57% had a normal BMI, and 39% were obese, of which, 13% were morbidly obese with a trend towards a higher BMI in later years. There was a significant association between a low BMI and female gender, history of diarrhea, reduced hemoglobin concentration, reduced bone mineral density, osteoporosis, and higher grades of villous atrophy. After treatment with the GFD for at least 2 years, weight gain was observed in 81% of patients, no change in only 4%, and weight loss in 15%. After treatment, 82% of initially overweight patients had gained more weight, and the proportion of patients in the overweight category increased from 26% to 51%. While weight gain in underweight patients is expected and desired following a GFD, this weight gain was also observed in patients with a normal BMI, of whom almost a third had entered the overweight category. The main cause of weight gain was more efficient intestinal absorption. Some patients may also be ingesting too many calories after diagnosis by eating comforts foods, such as candy or snacks that have been identified as safe, or by eating more meat, cheese, and other high-fat foods. Some gluten-free products are higher in fat and calories than their naturally gluten-free counterparts. Strategies for weight management include eating more low calorie foods, such as fresh fruits, vegetables, salads, and seafood; moderate amounts of low-fat dairy products, lean meats, and low-fat grain products; and limiting the intake of high-fat foods, sweets, and desserts. An increase in exercise can also counteract the tendency of weight gain.

An alteration in lipid metabolism following the initiation of a GFD was reported in recent studies [82]. In celiac patients under diet treatment, increases in the total cholesterol and high-density lipoprotein levels, and a decrease in the low-density/high-density lipoprotein ratio, was found. There are two possible mechanisms for these changes: increased intestinal synthesis of both high-density lipoprotein and apolipoprotein A-I due to the decreased intestinal production of apolipoprotein A-I during active CD, and increased fat absorption from the diet with the high high-density lipoprotein levels reflecting a change of the type of fat consumed.

A study designed to evaluate the vitamin nutrition status of a series of adult celiac patients adhering a strict GFD for 8-12 years found that half of the patients had a poor vitamin status [83]. To assess these deficiencies, the authors determined the total plasma homocysteine level, a metabolic marker of folate, vitamin B_6, and B_{12} deficiency and an independent risk factor for cardiovascular disease. This may have clinical implications considering the linkage between elevated total plasma homocysteine levels and cardiovascular disease, in the same range as hypercholesterolemia and hypertension. Evidence is emerging that an elevated total plasma homocysteine level may be a causal factor, and a dose response relationship has been found between high levels and mortality, particularly for a concentration at or above 15 μmol/L. Studies show that daily supplementation with 0.5-5 mg of folic acid or 0.5 mg of vitamin B_{12} can normalize the concentration of total plasma homocysteine [84].

The association between gastrointestinal disorders and impaired bone health has been known for many years. Undiagnosed and untreated CD constitutes an increasing skeletal health problem due to its association with low bone density, osteopenia, osteoporosis, osteomalacia, bone pain, and rickets, which means an increased fracture risk [85]. In children with classic CD, growth failure is a clinical manifestation in 70-80% of patients [86]. Normally, bone mass increases dramatically during growth and the amount of bone accrued during this period is an important determinant of future fracture risk. In CD patients, bone mass measurements are greatly reduced compared to healthy youth [87]. As indicated by surrogate markers of bone turnover, in CD, bone formation rates are depressed and bone resorption rates enhanced in untreated patients. Reduced calcium intake and impaired intestinal calcium absorption in the proximal small intestine have been considered for a long time to be the leading causes of low bone mineral density in CD. A subsequent elevation of parathyroid (PTH) hormone levels has thus been indicated as the possible cause of bone loss; however, the available data on bone metabolism indicates that the PTH-vitamin D axis is not affected in young celiac patients. More recent observations seem to indicate that imbalances of cytokines, such as tumor necrosis factor (TNF)-α, interleukin (IL)-1, and, specifically, IL-6, usually present in chronic inflammation and contribute to increased bone resorption with an up-regulation of osteoclast activity. Another possible pathogenetic mechanism for low bone mass in CD is related to the presence of autoimmunity against bone tissue transglutaminase [88].

Table 2. Modified from Kupper C. Dietary guidelines and implementation for Celiac Disease. *Gastroenterology,* **2005, 128: S121-S127**

At diagnosis	GFD	GF products	Long-term GFD
Calorie/protein			
Fiber	Fiber	Fiber	Fiber
Iron	Iron	Iron	
Calcium	Calcium		
Vitamin D	Vitamin D		
Magnesium	Magnesium		
Zinc			
Folate, niacin, B_{12}, riboflavin	Folate, niacin, B_{12}, riboflavin	Folate, thiamine, niacin, riboflavin	Folate, niacin, B_{12}, (w/supplements)

GFD, gluten-free diet; GF, gluten-free.

The measurement of bone density, serum calcium, alkaline phosphatase, and PTH hormone levels is thereby recommended at the time of diagnosis. Additional recommended measures include ensuring adequate calcium intake to maintain 1500 mg per day, supplementation with 400-800 IU of vitamin D, adhering to a strict GFD, exercise, abstaining from tobacco usage, minimizing alcohol intake, obtaining a baseline bone mineral density by dual energy-x-ray absorptiometry scan, and screening for vitamin D deficiency [14]. Some have recommended that bone mineral density measurements should be repeated 1 year following the GFD [89]. Dietary treatment alone is able to restore bone mass quite rapidly in the axial, peripheral, and whole skeleton. Attaining a complete catch-up of growth and histological recovery after adherence to a GFD has been documented. Bone mass improves when children follow a GFD. However, it has not been definitively established whether a GFD restores bone mass to normal values. The duration of a GFD has been found to be related to the improvement of bone mineral measurements; remarkable increases of bone mineral density necessitates at least one year of treatment.

THE GLYCEMIC INDEX (GI)

The concept of glycemic index (GI) was introduced by Jenkins and Wolever in 1981 as a quantitative indicator of the ability of carbohydrates to raise blood glucose. The GI is defined as the incremental area under the glucose response curve following the intake of 50 g of carbohydrates from a test food compared with the glucose area induced by the same amount of carbohydrate from a standard carbohydrate source. High-GI foods cause higher postprandial glycemia and insulinemia than low-GI foods. Carbohydrates, mainly starch from cereals, play an important part in a balanced diet, and dietary guidelines suggest a diet with low-GI foods [90]. Theoretically, the glycemic response of carbohydrates may be increased following the removal of gluten [91]; the gluten protein network surrounds the starch granules, not allowing amylase to easily access the granule and inhibiting the rate of starch hydrolysis in the lumen of the small intestine.

Since CD is associated with a high incidence of type 1 diabetes [92], an important task for these patients is to select foods with both beneficial effects on lipid metabolism and minimal hyperglycemic activity while adhering to a strict GFD. In light of the knowledge that insulin resistance and a raised postprandial triglyceride concentration are risk factors for cardiovascular disease, the objective for the dietary treatment of individuals with CD and diabetes should be to achieve good glycemic control while maintaining a strict GFD [93].

FUTURE THERAPEUTIC OPTIONS

The complexity of a GFD, the number of patients with truly refractory disease or the patients unable to tolerate the diet, and the growing awareness of the high prevalence of CD in the general population highlight the necessity for advancements in the treatment of CD. The future therapeutic opportunities include modifications of gluten processing, targeting of HLA-associated molecular interactions, oral therapy with prolyl endopeptidases, and cytokine regulation.

The identification of epitopes on α-gliadin that function in T-cell activation has led to the hypothesis that the elimination or modification of specific gliadin epitopes via the genetic modification of wheat proteins may lead to safer, less immunogenic, wheat-containing commercial products. However, genetically modifying food is controversial and, though this process is theoretically enticing, it may prove more complicated than previously thought [94,95]. Alternative strategies include the development of gliadin analogs that are capable of interfering with HLA binding and subsequent T-cell activation, or the alternative delivery of dominant gliadin epitopes to develop tolerance to glutens. The kind of strategies proposed include intranasal delivery of whole gluten or specific epitopes, DNA vaccine technology with plasmids to specific gluten epitopes, antigen-presenting cell genetic engineering to present gluten peptides, and the use of FAS ligand to target FAS-positive T cells for the induction of apoptosis and to block the CD28/B7 co-stimulatory pathway. While this latter strategy has many potential benefits, it does have the risk of enhancing immunization and CD pathogenesis rather than promoting tolerance [96].

An up-regulation of the prolyl endopeptidases (PREP) gene and PREP activity has been demonstrated in CD patients [97,98]. Recently, some authors suggested the use of bacteria-derived prolyl endoproteases to cleave gliadin peptides, resulting in the decreased delivery of these peptides for epithelial processing, which is the more integral component in CD pathogenesis. Furthermore, the delivery of encapsulated endoproteases is a relatively straightforward and testable strategy that may eliminate the need for a GFD in the future [99]. The three prolyl endopeptidases tested were from different bacterial sources: *Flavobacterium meningosepticum*, *Saphingomonas capsulate*, and *Myxococcus xanthus*.

A mixture of four sourdough lactobacilli has also been demonstrated to possess the ability to hydrolyze gliadin fractions after prolonged incubation (12-24 hours) with dough made from a combination of flours, including wheat. Bread baked using this dough failed to demonstrate any increase in intestinal permeability in celiac patients, in contrast to bread baked using standard baker's yeast, which did induce a marked alteration of intestinal permeability [100].

The increasing knowledge of specific cytokines involved in CD pathogenesis provides further potential therapeutic options [101]. Particularly, alternative methods of delivery or regulation of IL-10 and IL-15 have been proposed. The expression of IL-10 has been proposed as a target for potential modulation based on similarities between the immune responses in CD and Crohn's disease. Interleukin-10 inhibits interferon (IFN)γ-producing T cells by preventing macrophage and dendritic cells from synthesizing IL-12 with a protective function against inflammation. In CD, IL-10 release is up-regulated upon T-cell activation. Mice deficient in IL-10 spontaneously develop enterocolitis mediated by CD4+ T cells. It is possible that a further increase in IL-10 expression in CD will result in a blunted immunological response to gluten [102-104].

Interleukin-15, which is produced by macrophages, appears to play an integral role in bridging the innate and adaptive immune responses in CD. For example, IL-15 induces direct epithelial injury via mechanisms relating to IFNγ, enhances the capacity of dendritic cells to function as antigen-presenting cells, and regulates intraepithelial lymphocyte homeostasis by promoting migration and preventing apoptosis. For these reasons, the neutralization of IL-15 expression or function is a prime target for future therapies, possibly with a specific IL-15 inhibitor, or IL-15 neutralization via IL-15 receptor inhibition [105].

REFERENCES

[1] Hoffenberg E.J., Mackenzie T. et al. (2003). A prospective study of the incidence of childhood celiac disease. *J Pediatr, 143*,308–314.

[2] Fasano A., Berti I. et al. (2003). Prevalence of celiac disease in at-risk and not-at-risk groups in the United States: a large multicenter study. *Arch Intern Med, 163,* 286-292.
[3] Sollid LM. (2000). Molecular basis of celiac disease. *Annu Rev Immunol, 18,* 53-81.
[4] Green P.H., Jabri B. (2006). Celiac disease. *Annu Rev Immunol, 57,* 207-21.
[5] Farrell R.J., Kelly C.P. (2002). Celiac sprue. *N Engl J Med, 346(3),* 180-8.
[6] Fasano A., Catassi C. (2001). Current approaches to diagnosis and treatment of celiac disease: an evolving spectrum. *Gastroenterology, 120,* 636-51.
[7] Green P.H.R., Stavropoulos S.N. et al. (2001). Characteristics of adult celiac disease in the USA: results of a national survey. *Am J Gastroenterol, 96(1),* 126-31.
[8] Peters U, Askling J. et al. (2003). Causes of death in patients with celiac disease in a population-based Swedish cohort. *Arch Intern Med, 163(13),* 1566-72.
[9] West J., Logan R.F. et al. (2004). Malignancy and mortality in people with coeliac disease: population based cohort study. *BMJ, 329(7468),* 716-9.
[10] Corrao G., Corazza G.R. et al. (2001). Mortalità in patients with coeliac disease and their relatives: a cohort study. *Lancet, 358(9279),* 356-61
[11] Askling J., Linet M. et al. (2002). Cancer incidence in a population-based cohort of individuals hospitalized with celiac disease dermatitis herpetiformis. *Gastroenterology, 123(5),* 1428-35.
[12] Kemppaienen T., Kroger H. et al. (1999). Osteoporosis in adult patients with celiac disease. *Bone, 24,* 249-255.
[13] American Gastroenterological Association. (2001). American Gastroenterological Association medical position statement: celiac sprue. *Gastroenterology, 120,* 636-651.
[14] Scott E.M., Gaywood I. et al. (2000). Guidelines for osteoporosis in celiac disease and inflammatory bowel disease. British Society of Gastroenterology. *Gut, 46(suppl 1),* 1-8.
[15] Mora S., Barera G. et al. (2001). A prospective, longitudinal study of the long-term effect of treatment on bone density in children with celiac disease. *J Pediatr, 139,* 516-521.
[16] Rostami K., Steegers E.A.P. et al. (2001). Coeliac disease and reproductive disorders: a neglected association. *Eur J Obstet Gynecol Reprod Biol, 96(2),* 146-9.
[17] Smecuol E., Maurino E. et al. (1996). Gynaecological and obstertric disorders in coeliac disease: frequent clinical onset during pregnancy or the puerperium. *Eur J Gastroenterol Hepatol, 8,* 63-9.
[18] Martinelli P., Troncone R. et al. (2000). Coeliac disease and unfavourable outcome of pregnancy. *Gut, 46,* 332-5
[19] Meloni G.F., Dessole S. et al. (1999). The prevalence of coeliac disease in infertility. *Hum Reprod, 14,* 2759-61.
[20] See J., Murray J.A. (2006). Gluten-free diet: the medical and nutrition management of celiac disease. *Nutr Clin Pract, 21(1),* 1-15.
[21] Ciacci C., Iavarone M. et al. (2002). Psychological dimensions of celiac disease. *Dig Dis Sci, 47,* 2082-2087.
[22] Hallert C., Granno C. et al. (2002). Living with celiac disease: controlled study of the burden of illness. *Scand J Gastroenterol, 37,* 39-42.
[23] Murray J.A. (1999). The widening spectrum of celiac disease. *Am J Clin Nutr, 69,* 354-365.
[24] American Dietetic Association. (2000). *Manual of Clinical Dietetics.* (6th ed). Chicago, IL: American Dietetic Association.

[25] Celiac Disease Foundation. (2002). *Guidelines for a gluten free lifestyle.* (3rd ed). Studio City, CA: Celiac Disease Foundation.
[26] Case S. (2002). *Gluten free diet: a comprehensive resource guide.* (2nd ed). Regina Sashatchewan: Case Nutrition Consulting.
[27] Crowe J.P., Falini N.P. (2001). Gluten in pharmaceutical products. *Am J Health Syst Pharm, 58,* 396-401.
[28] Hill I.D., Dirks M.H. et al. (2005). Guideline for the diagnosis and treatment of celiac disease in children: recommendations of the North American Society for Pediatric Gastroenterology, Hepatology and Nutrition. *J Pediatr Gastroenterol Nutr, 40,* 1-19.
[29] Liu S., Sesso H.D., et al. (2003). Is intake of breakfast cereals related to total and cause specific mortality in men? *Am J Clin Nutr, 77,* 594-599.
[30] Montonen J., Knekt P., et al. (2003). Whole-grain and fiber intake and the incidence of type 2 diabetes. *Am J Clin Nutr, 77,* 622-629.
[31] Slavin J.L., Martini M.C. et al. (1999). Plausibile mechanisms for the protectiveness of whole grains. *Am J Clin Nutr, 70 (suppl),* 459S-463S.
[32] Murray J.A., Watson T. et al. (2004). Effects of a gluten-free diet on gastrointestinal symptoms in celiac disease. *Am J Clin Nutr, 79,* 669-673.
[33] Abdulkarim A.S., Burgart L.F., et al. (2002). Etiology of nonresponsive celiac disease: results of a systematic approach. *Am J Gastroenterol, 97,* 2016-2021.
[34] Pourpak Z., Mesdaghi M., et al. (2005). Which cereal is a suitable substitute for wheat in children with wheat allergy? *Paediatr Allergy Immunol, 16,* 262-6
[35] Vader L.W., de Ru A., et al. (2002). Specificity of tissue transglutaminase explains cereal toxicity in coeliac disease. *J Exp Med, 195,* 643-9.
[36] Janatuinen E.K., Pikkarainen P.H., et al. (1995). A comparison of diets with or without oats in adult with celiac disease. *N Engl J Med, 333,* 1033-7
[37] Arentz-Hansen H., Fleckenstein B. et al. (2004). The molecular basis for oat intolerance in patients with celiac disease. *Plos Medicine, 1.* 084-092.
[38] Hogberg L., Laurin P. et al. (2004). Oats to children with newly diagnosed coeliac disease: a randomised double blind study. *Gut, 53.* 649-654.
[39] Holm K., Maki M. et al. (2006). Oats in the treatment of childhood coeliac disease: a 2-year controlled trial and a long-term clinical follow-up study. *Aliment Pharmacol Ther, 23,* 1463-1472.
[40] Catassi C, Rossini M et al. (1993). Dose dependent effects of protracted ingestion of small amounts of gliadin in coeliac disease children: a clinical and jejunal morphometric study. *Gut, 34.* 1515-19.
[41] Peraaho M., Mustalathi K. et al. (2004). Effect of an oats-containing gluten-free diet on symptoms and quality of life in coeliac disease. A randomized study. *Scand J Gastroenterol, 39,* 27-31.
[42] Storsrud S., Olsson M., et al. (2003). Adult coeliac patients do tolerate large amounts of oats. *EUr J Clin Nutr, 57,* 163-9.
[43] Haboubi N.Y., Taylor S. et al. (2006). Coeliac disease and oats: a systematic review. *Postgrad Med J, 82.* 672-678.
[44] Calaco J., Egan-Mitchell B., et al. (1987). Compliance with gluten-free diet in coeliac disease. *Arch Dis Child, 62,* 706-8
[45] Bardella M.T., Molteni N., et al. (1994). Need for follow-up in coeliac disease. *Arch Dis Child, 70,* 211-3

[46] Mariani P, Viti M.G., et al. (1998). The gluten-free diet: a nutritional risk factor for adolescent with celiac disease? *J Pediatr Gastroenterol Nutr27(5)*,519-23

[47] Fabiani E., Taccari L.M., et al. (2000). Compliance with gluten-free diet in adolescents with screening-detected celiac disease: a 5 year follow-up. *J Pediatr, 136,* 841-3.

[48] Lee A., Newman J.M. et al. (2003). Celiac disease: its impact on quality of life. *J Am Diet Assoc, 103,* 1533-1535.

[49] Casellas F., Rodrigo L., et al. (2008). Factors that impact health-related quality of life in adults with celiac disease: a multicenter study. *World J Gastroenterol, 14(1),*46-52.

[50] Hallert C., Granno C., et al. (1998). Quality of life of adults celiac patients treated for ten years. *Scand J Gastroenterol, 9,* 933-938.

[51] Joint FAO/WHO. (1981) Food Standards program Codex alimentarius Commission. Codex stan

[52] Skerritt JH., Devery JM., et al. (1990). Gluten intolerance: Chemistry, celiac-toxicity, and detection of prolamins in foods. *Cereal Foods world, 35,* 638-44.

[53] Stuart ML., Faulkner-Hogg KB., et al. (1997). Dietary gluten restriction and symptom occurrence in a large population of subjects with celiac disease. *J Gastroenterol Hepatol, 12 Suppl*: A42

[54] Ciclitira PJ., Cerio R. et al. (1985). Evaluation of a gliadin containing gluten-free product in coeliac patients. *Hum Nutr Clin Nutr, 39C,* 303-308.

[55] Chartland LJ., Russo PA., et al. (1997). Wheat starch intolerance in patients with celiac disease. *J Am Dent assoc, 97,* 612-618.

[56] Maki M., Collin P. (1997). Coeliac disease. *Lancet, 349,* 1755-1759

[57] Ciclitira PJ, Evans DJ, et al. (1984). Clinical testing of gliadin fractions in coeiac patients. *Clin sci, 66,* 357-64

[58] Ejderhamn J., Veress B., et al. (1988). The long term effect of continual ingestion of wheat starch –containing gluten-free products in celiac patients. In: Kumar PJ, ed. *Coeliac disease: one hundred years*. Leeds, united Kingdom: Ledds University Press, 294-297.

[59] Kaukinen K., Collin P., et al. (1999). Wheat starch-containing gluten free flour products in the treatment of celiac disease and dermatitis herpetiformis. A long term follow-up study. *Scand J Gastroenterol, 34,* 909-914

[60] Catassi C., Fabiani E., et al. (2007). A prospective, double-blind, placebo-controlled trial to establish a safe gluten treshold for patients with celiac disease. *Am J Clin Nutr, 85,* 160-166

[61] Hopman E.G.D., le Cessie S., et al. (2006). Nutritional management of the gluten free diet in young people with celiac disease in the Netherlands. *J Pediatr Gastroenterol Nutr,43(1),*102-8.

[62] Halfdanarson T.R., Litzow M.R., et al. (2006). Hematological manifestations of celiac disease. *Blood, 109(2),* 412-21.

[63] Mant M.J., Bain V.G., et al. (2006). Prevalence of occult gastrointestinal bleeding in celiac disease. *Clin Gastroenterol Hepatol, 4(4),*451-4

[64] Dahele A., Ghosh S. (2001). Vitamin B12 deficiency in untreated celiac disease. *Am J Gastroenterol, 96,* 745-750.

[65] Bode S., Gudmand-Hoyer E. (1996). Symptoms and haematologic features in consecutive adult celiac patients. *Scand J Gastroenterol, 31(1),* 54-60.

[66] Clifford L. (2006). Micronutrient deficiencies. In: Buchman AL, editor. Clinical nutrition in gastrointestinal disease. Thorofare (NJ): SLACK Inc.
[67] Holick M.F. (2003). Vitamin D deficiency: what a pain it is. *Mayo Clin Proc, 78(12),*1457-9.
[68] Kozanoglu E., Basaran S., et al. (2005). Proxymal myophaty as an unusual presenting feature of celiac disease. *Clin Rheumatol, 24(1),* 76-8
[69] Howard M.R., Turnbull A.J. et al. (2002). A prospective study of the prevalence of undiagnosed celiac disease in laboratory defined iron and folate deficiency. *J CLin Pathol, 55(10),* 754-7
[70] Haapalahti M., Kulmala P, et al. (2005). Nutritional status in adolescents and young adults with screen-detected celiac disease. *J Pediatr Gastroenterol Nutr, 40(5),* 566-70.
[71] Prasad A.S. (1998). Zinc and immunity. *Mol Cell Biochem,188,* 63-9
[72] Bedwal RSBA. (1994). Zinc, copper and selenium in reproduction. *Experimentia, 50,* 626-40.
[73] Bougle D., Proust A. (1999). Iron and zinc supplementation during pregnancy: interactions and requirements. *Contracept Fertil Steril, 27,* 537-43.
[74] Vezina D., Mauffette F., et al. (1996). Selenium-vitamin E supplementation in infertile men. Effects on semen parameters and micronutrient levels and distribution. *Biol Trace Elem Res, 53,* 65-83
[75] Goyens P., Brasseur D., et al. (1985). Copper deficiency in infants with active celiac disease. *J Pediatr Gastroenterol Nutr, 4(4),* 677-80.
[76] Jameson S., Hellsing K., et al. (1985). Copper malabsorption in coeliac disease. *Sci Total Environ, 42(1-2),* 29-36.
[77] Goldman A.S., van Fossen D.D., et al. (1962). Magnesium deficiency in coeliac disease. *Pediatrics, 29,* 948-52.
[78] Rujner J., Socha J., et al. (2004). Magnesium status in children and adolescents with coeliac disease without malabsorption symptoms. *Clin Nutr, 23(5),*1074-9.
[79] Viljamaa M., Collin P., et al. (2005) Is coeliac disease screening in risk group justified ? A fourteen-year follow-up with special focus on compliance and quality of life. *Aliment Pharmacol Ther, 20,* 317-24.
[80] West J., Logan R.F.A. (2004). Risk of vascular disease in adults with diagnosed coeliac disease: a population based study. *Aliment Pharmacol Ther, 20,* 73-79.
[81] Dickey W., Kearney N. (2006). Overweight in celiac disease: prevalence, clinical characteristics and effect of a gluten-free diet. *Am J Gastroenterol, 101(10),* 2356-9.
[82] Brar P., Kwon G.Y., et al. (2006). Change in lipid profile in celiac disease: beneficial effect of glute-free diet. *Am J Med, 119(9),* 786-90.
[83] Hallert C., Grant C., et al. (2002). Evidence of poor vitamin status in coeliac patients on a gluten free diet for 10 years. *Aliment Pharmacol Ther, 16,* 1333-1339.
[84] Homocysteine Lowering Trialists Collaboration. (1998). Lowering blood homocysteine with folic acid based supplements: metanalysis of randomised trials. *Br Med J, 316,* 894-8.
[85] Cranney A., Rostom A., et al. (2005). Consequences of testing fo celiac disease. *Gastroenterology,128,* 109-120.
[86] George E.K., Mearin M.L., et al. (1997). Twenty years of childhood celiac disease in the Netherlands: a rapidly increasing incidence. *Gut, 40,*61-66.

[87] Mora S., Gilsanz V. (2003). Establishment of peak bone mass. *Endocrinol Metab Cliin N Am. 32*, 39-63.
[88] Corazza G.R., Di Sario A., et al. (1995). Bone mass and metabolism in patients with celiac disease. *Gastroenterology, 109*, 122-128.
[89] Corazza G.R., Di Stefano M., et al. (2005). Bones in coeliac disease: diagnosis and treatment. *Best Pract Res Clin Gastroenterol, 19(3)*, 453-65.
[90] Brand CJ., Nicholson PL. et al. (1985). Food processing and the glycemic index. *Am J Clin Nutr, 42*, 1192-1196.
[91] Jenkins DJA., Thorne MJ., et al. (1987). The effect of starch-protein interaction in wheat on the glycemic response and rate in vitro digestion. *Am J Clin Nutr, 45*, 946-951
[92] Cronin CC., Shanahan F., et al. (1997). Insulin-dependent diabetes mellitus and coeliac disease. *Lancet, 349*, 1096-1097
[93] Jenkins DJ., Thorne MJ., et al. (1987). The effect of the starch-protein interaction in wheat on the glycemic response and rate of in vitro digestion. *Am J Clin Nutr, 62*, 946-951
[94] Vader W., Kooy Y., et al. (2002). The gluten response in children with celiac disease is directed toward multiple gliadin and glutenin peptides. *Gatroenterology, 122*, 1729-37
[95] Vanderlugt C.J., Miller S.D. (1996). Epitope spreading. *Curr Opin Immunol, 8*, 831-6.
[96] Martucci S., Corazza G.R. (2002). Spreading and focusing of gluten epitopes in celiac disease. *Gastroenterology, 122*, 2072-5.
[97] Diosdado B., Wapenaar M.C., et al. (2004). A microarray screen for novel candidate genes in coeliac disease pathogenesis.. *Gut, 53*, 944-951
[98] Matysiak-Budnik T., Candalh C., et al. (2003). Alterations of the intestinal transport and processing of gliadin peptides in celiac disease. *Gastroenterology, 125*, 696-707.
[99] Shan L., Molberg O., et al. (2002). Structural basis for gluten intolerance in celiac sprue. *Science, 297*, 2275-9.
[100] Di Cagno R., De Angelis M., et al. (2004). Sourdough bread made from wheat and non-toxic flours and started with selected lactobacilli is tolerated n celiac sprue patients. *Appl Environ Microbiol, 70*, 1088-1096.
[101] Londei M., Quarantino S., et al. (2003). Celiac disease: a model autoimmune disease with gene therapy applications. *Gene Ther, 10*, 835-43.
[102] D'Andrea A., Aste-Amezaga M., et al. (1993). Interleukin 10 (IL-10) inhibits human lymohicyte interferone gamma-production by suppressing natural killer cell stimutaroy factor/IL-12 synthesis in accessori cells. *J Exp Med, 178*, 1041-8.
[103] Davidson N.J., Leach M.W., et al. (1996). T helper cell 1-type CD4+ T cells, but not B cells, mediate colitis in interleukin 10 deficient mice. *J Exp Med, 184*, 241-51.
[104] Van Montfrans C., Rodriguez Pena M.S., et al. (2002). Prevention of colitis by interleukin 10-transduced T lymphocytes in the SCID mice transfer model. *Gastroenterology, 123*, 1865-76.
[105] Mention J.J., Ben Ahmed M., et al. (2003). Interleukin 15: a key to disrupted intraepithelial lymphocyte homeostasis and lymphomagenesis in celiac disease. *Gastroenterology, 125*, 730-45.

INDEX

A

abnormalities, 172
abortion, 172
absorption, 171, 172, 173, 174
acetic acid, 93, 95
acid, ix, x, 3, 12, 15, 19, 29, 32, 34, 35, 43, 45, 51, 53, 69, 80, 93, 95, 96, 103, 108, 113, 115, 116, 119, 126, 129, 130, 133, 135, 136, 137, 139, 140, 141, 142, 143, 144, 145, 146, 147, 148, 149, 150, 152, 153, 157, 159, 162, 171, 172, 174, 180
acidic, 19, 103
acquired immunity, 109
activation, 162, 175, 176
active oxygen, 44
actuators, viii, 28
adaptability, 121
additives, 3, 5, 10, 11, 15, 164
adenocarcinoma, 163
adolescents, xi, 161, 169, 171, 179, 180
adsorption, 115, 121, 134, 147, 158
adult, 162, 167, 173, 177, 178, 179
adult population, 162
adults, xi, 7, 162, 167, 168, 170, 179, 180
advancements, 175
adverse effects, 7, 30, 77
aggregates, 145
aggregation, ix, 5, 36, 39, 89, 90, 92, 93, 94, 95, 96, 97, 114, 115, 141
alanine, 95
albumin, 4, 11, 36, 153, 158
alcohol, 163, 174
alkaline, x, 133, 135, 137, 141, 144, 145, 146, 147, 148, 149, 150, 152, 153, 163, 174
alkaline hydrolysis, 141
alkaline phosphatase, 163, 174
allergens, 78, 79
allergic reaction, viii, 49

allergy, 2, 6, 86, 178
alopecia, 172
alternative, 135, 164, 175, 176
alternatives, x, 133, 134
amelioration, 167
amenorrhea, 163
amine, 137, 142
amine group, 137, 142
amino, vii, x, 10, 27, 28, 29, 30, 32, 34, 35, 40, 43, 44, 45, 51, 69, 80, 91, 93, 95, 103, 107, 108, 116, 133, 141, 142, 143, 145, 155, 162
amino acid, vii, 10, 27, 28, 29, 30, 32, 34, 35, 40, 43, 44, 45, 51, 69, 80, 91, 93, 95, 103, 107, 108, 141, 143, 145, 155, 162
amino acid side chains, 141
amino acids, vii, 10, 27, 28, 30, 34, 35, 40, 43, 44, 45, 51, 91, 93, 95, 103, 107, 108, 145, 155
amino groups, x, 133, 142
ammonia, 30
ammonium, 30, 31, 40, 42, 46
amylase, 175
anaphylactic shock, 6
anaphylaxis, 6, 23
anemia, 85, 170, 171, 172
ANOVA, x, 113, 120, 126, 128
antibody, 6, 80, 86, 87, 165
antigen, ix, 6, 76, 101, 104, 107, 108, 109, 111, 176
antigen-presenting cell, 104, 107, 176
anti-gliadin, ix, 101
antioxidant, 15, 18, 70
antioxidative activity, 52
anti-transglutaminase antibodies, ix, 101
anxiety, 164, 169
apolipoprotein A-I, 173
apoptosis, 109, 110, 176
appetite, 173
Argentina, 113, 133
arthritis, 162
ascorbic acid, 3, 43

ash, 135
assessment, 24, 52, 78, 163, 167, 170
assimilation, 30, 33, 40, 44, 46, 47
asymptomatic, 7, 162
atmosphere, 28, 33, 69
ATP, 33
atrophy, ix, 6, 101, 102, 106, 107, 162, 167, 168, 173
attachment, 106
Australia, 170
authority, xi, 161
autoantibodies, ix, 101, 102, 106, 110, 112
autoantigens, 106
autoimmune, 163, 181
autoimmune disease, 181
autoimmune diseases, 111
autoimmune disorders, 163
autoimmune enteropathy, viii, 75
autoimmunity, 6, 102, 107, 111, 112, 174
availability, 145, 164, 167
awareness, 175

body weight, 163
bonding, ix, 5, 66, 89, 90, 92, 97
bonds, viii, 4, 5, 12, 14, 28, 29, 36, 38, 39, 43, 45, 66, 89, 90, 91, 92, 93, 94, 95, 96, 103, 137, 143, 153, 158
bone, xi, 162, 163, 164, 173, 174, 177, 181
bone density, 163, 174, 177
bone form, 174
bone loss, 174
bone mass, 174, 175, 181
bone resorption, 174
bones, 20
bowel, 87, 102, 105, 111, 162, 167, 168, 171, 177
Brazil, 1, 6, 18, 19, 51
buckwheat, viii, 9, 16, 18, 23, 25, 49, 51, 52, 53, 54, 55, 57, 58, 59, 60, 61, 62, 63, 64, 66, 67, 68, 69, 70, 71, 72
buffer, 137
buns, 3
by-products, ix, x, 113, 114, 133, 134

B

B vitamins, 166
Bacillus, 135
backscattering, 131
bacteria, 105, 176
bacterial, 167, 176
baking quality, vii, viii, 27, 28, 29, 32, 33, 36, 37, 38, 39, 40, 47, 90, 91, 97, 100
barley, x, 161, 162, 167
base, ix, 12, 14, 57, 67, 75, 80, 81, 101, 118
beer, 80, 164
behavior, 134, 138, 141, 147, 150, 152, 158, 159
Belgium, 19, 53
beneficial effect, 175, 180
benefits, viii, 3, 17, 31, 49, 176
benzoyl peroxide, 3
beverages, 68, 71, 72, 85
bicarbonate, 53, 55
binding, 175
biological roles, 45
biopolymers, 52
biopsy, xi, 6, 104, 161, 168
biosynthesis, 43, 45
birth weight, 162
bleeding, 171, 179
blends, 52, 53, 54, 55, 61, 65, 67
blindness, 172
blood, 6, 104, 111, 171, 172, 175, 180
blood glucose, 175
blood smear, 172
BMI, 173
body mass index, 173

C

Ca^{2+}, 70, 106
calcium, xi, 20, 52, 72, 108, 161, 162, 163, 164, 166, 171, 172, 174
caloric intake, 165
calorie, 173
cancer, 162, 166
carbohydrate, 175
carbohydrates, xi, 2, 22, 51, 161, 171, 175
carbon, 42
carboxymethyl cellulose, 72
carcinoma, 110
carcinomas, 163
cardiovascular disease, xi, 70, 162, 173, 175
casein, 130, 158
CD8+, 106, 109
Celiac disease, x, 161, 162, 177, 179, 181
Celiac disease (CD), viii, x, 75, 161, 162
celiac patients, vii, 1, 2, 8, 10, 11, 13, 14, 17, 50, 52, 76, 85, 104, 110, 162, 163, 164, 168, 169, 171, 172, 173, 174, 176, 179
celiac sprue, 102, 177, 181
cell, 162, 168, 172, 175, 176, 181
cell division, 172
cell line, 104
cell lines, 104
cellulose, 12, 15, 16, 23, 53, 59, 72
cereals, 164, 167, 168, 175, 178
cheese, 14, 166, 173
chemical, vii, 1, 5, 8, 12, 16, 18, 62, 68, 98, 132, 159
chemical characteristics, 62
chemical properties, 8, 98

chemiluminescence, 80
Chicago, 130, 132, 158, 177
childhood, xi, 21, 105, 161, 176, 178, 180
children, vii, 6, 87, 105, 164, 166, 167, 168, 174, 175, 177, 178, 180, 181
China, 51, 70
cholesterol, 51, 173
chromatograms, 140, 144, 146, 147, 148, 149, 151, 152, 153
chromatography, 37, 54, 79, 81, 94, 99, 159
chromosome, 36, 104
chronic disorders, 162
classes, 69, 93
classification, 71
cleavage, 39, 93, 114, 137, 139, 150, 151, 153, 158
clusters, 105
CMC, 11, 51, 53, 55, 56, 65, 66, 68
CO_2, 40, 51
coeliac disease, 177, 178, 180, 181
combined effect, 24
commercial, ix, x, 11, 14, 55, 79, 84, 87, 94, 95, 113, 133, 164, 168, 175
compatibility, 6, 8
competence, 165
complement, 81
complexity, 111, 122, 163, 175
compliance, xi, 15, 23, 110, 161, 163, 164, 165, 169, 170, 171, 180
complications, 76, 109, 162, 163, 168, 169
components, 134, 140, 141, 142, 143, 144, 146, 147, 148, 149, 150, 151, 152, 153, 155
composition, vii, viii, ix, x, 18, 27, 28, 29, 31, 32, 33, 36, 37, 38, 42, 44, 45, 46, 47, 51, 68, 82, 89, 93, 97, 99, 102, 103, 115, 129, 134, 135, 143, 145, 157
compounds, 15, 16, 22, 34, 45, 64, 72, 81
comprehension, 165
compression, 5, 57, 64
concentration, x, 133, 135, 136, 139, 145, 148, 150, 153, 163, 172, 173, 175
concordance, 104
conductivity, 137
conference, 20
configuration, 95, 104
congenital malformations, 172
constituents, 16, 42, 63
consumers, 14, 78
consumption, vii, viii, xi, 1, 2, 3, 7, 49, 52, 161, 168, 169, 170, 171
contaminant, 80
contaminated food, 82, 83
contamination, xi, 7, 77, 80, 87, 161, 164, 167, 168
contradiction, 61

control, 171, 175
controlled trials, 78
controversial, 37, 108, 167, 170, 175
cooking, 12
cooling, 8, 10, 57
copolymers, 131
correlation, 32, 36, 39, 45, 91, 147, 149, 150, 153, 155, 157, 171
correlations, 144, 146, 155
corticosteroids, 167
cost, 8, 17, 111, 114, 134, 167
counsel, 163
counseling, 163, 164
counterbalance, 171
covalent bond, 12, 94, 103
coverage, 134
creep, 13, 15, 23
crop, 8, 28
crops, 16, 44, 51
cross-linking, 167
crust, 9, 12, 13, 14, 16, 64, 65, 68, 96
crypt hyperplasia, ix, 6, 101, 102, 107, 162
cultivars, 28, 31, 32, 42, 80, 86, 96, 100
cultivation, 35, 102
cycling, 44
cysteine, vii, 27, 29, 34, 35, 36, 38, 39, 40, 43, 45, 93, 94, 96, 103
cytokine, 175
cytokines, ix, 101, 109, 174, 176
cytoskeleton, 110
cytotoxicity, 107, 109, 110

D

dairy, 166, 173
dairy products, 166, 173
decisions, 169
defects, 15
deficiencies, xi, 162, 163, 172, 173, 180
deficiency, viii, 28, 31, 35, 36, 37, 38, 39, 40, 45, 46, 52, 85, 166, 167, 171, 172, 173, 174, 179, 180
definition, xi, 161
denaturation, ix, 70, 89, 90, 92, 93, 95, 97, 99, 131, 158
dendritic cell, 109, 176
density, xi, 162, 163, 173, 174, 177
Department of Agriculture, 24
Department of Health and Human Services, 86
depolymerization, 39
depressed, 174
depression, 164, 169
derivatives, 15, 23, 76, 102, 164
dermatitis, 7, 106, 177, 179
dermatitis herpetiformis, 7, 106, 177, 179

desiccation, 38
desorption, 81, 85, 129
destruction, 52
detectable, 79, 170, 171
detection, 54, 79, 80, 81, 82, 83, 84, 87, 170, 179
detoxification, 34, 42, 110
developing countries, 8
deviation, 63, 64, 66, 117, 120
diabetes, 162, 166, 175, 178, 181
diet, vii, viii, ix, xi, 1, 2, 6, 7, 8, 10, 13, 22, 35, 50, 52, 70, 71, 72, 75, 85, 87, 101, 102, 104, 105, 106, 107, 110, 161, 162, 163, 165, 167, 168, 169, 170, 171, 173, 174, 175, 177, 178, 179, 180
dietary, 162, 163, 165, 167, 169, 170, 171, 173, 175
dietary fiber, 52, 72
dietary habits, 171
dietary intake, 167
diffusion, 121, 134
digestibility, 7, 35, 45, 114
digestion, 17, 76, 111, 130, 157, 181
diseases, 2, 104, 111, 163, 170
disorder, ix, 6, 101, 102, 111, 112
dispersion, 117
distribution, x, 43, 46, 64, 76, 91, 96, 97, 100, 113, 114, 115, 117, 119, 120, 121, 122, 130, 159, 180
disulfide bonds, viii, 4, 5, 28, 29, 36, 38, 39, 43, 153, 158
diversification, ix, x, 113, 133
diversity, 167
division, 172
DNA, 81, 82, 83, 172, 176
DNA polymerase, 81
double blind study, 178
dough, vii, viii, 4, 5, 6, 9, 10, 11, 12, 13, 14, 15, 16, 18, 20, 21, 22, 23, 24, 25, 27, 28, 32, 33, 36, 38, 39, 40, 42, 43, 45, 46, 49, 50, 52, 53, 55, 56, 58, 59, 62, 63, 64, 65, 66, 67, 68, 70, 72, 76, 77, 90, 91, 97, 98, 99, 100, 103, 176
drainage, x, 133, 138, 139, 149, 150, 151, 152, 153, 154, 155, 156
drawing, 13
droplet size distribution, x, 113, 115, 120

E

ear emergence, vii, 27, 31, 35, 37, 38
eating, xi, 161, 163, 164, 168, 173
education, 163, 164
egg, 10, 11, 13, 14, 15, 51, 72, 130
electric circuits, 54
electrodes, 54
electron, viii, 49, 52, 54, 67, 99
electron microscopy, 52, 54, 67, 99
electron-microscopic analysis, viii, 49

electrophoresis, 32, 37, 38, 44, 52, 54, 80, 81, 90, 96, 99
ELISA, 76, 79, 80, 81, 82, 83, 84, 86
elucidation, 29
emission, 46
emotional, 164, 169
emotional well-being, 169
emulsifying, 158, 159
emulsions, x, 52, 69, 70, 113, 115, 116, 117, 118, 119, 120, 122, 123, 124, 125, 126, 127, 128, 129, 130, 131, 132, 134, 158
endosperm, 5, 18, 42, 44, 59, 62, 63
energy, 2, 11, 63, 92, 114, 134, 171, 174
engineering, 176
entanglements, 66
enterocolitis, 176
environment, vii, ix, 27, 47, 77, 89, 99
environmental factors, 21
enzymatic activity, 106
enzyme, ix, x, 13, 19, 20, 24, 76, 79, 80, 87, 102, 106, 108, 110, 113, 114, 115, 116, 129, 134, 135, 139, 141, 143, 147, 149, 155, 157, 167
enzyme immunoassay, 87
enzyme-linked immunosorbent assay, 76, 79, 87
enzymes, ix, x, 3, 5, 7, 10, 13, 17, 20, 51, 70, 111, 113, 115, 119, 133, 135, 137, 138, 141, 142, 143, 151, 172
enzymic hydrolysis, ix, x, 113, 131, 133, 159
epidemiology, 85
epithelial cells, 106, 109
epithelial membrane, ix, 101
epithelium, 7, 102, 105, 107, 108, 110, 162
epitopes, 168, 175, 181
equipment, 9, 77, 115, 117, 119, 122, 127
erythropoietin, 172
ester, 18
ethanol, 29, 76, 82, 83, 93
Europe, 7, 46, 51, 75, 86
evidence, ix, xi, 90, 93, 101, 102, 105, 107, 161, 163, 167
exclusion, 7, 81, 159
exercise, 6, 23, 173, 174
experimental condition, 119
expertise, 163
exposure, 78, 90, 97, 102, 111, 114, 162
extraction, viii, x, 4, 54, 62, 75, 82, 83, 84, 86, 114, 129, 135, 150, 157
extracts, x, 81, 84, 96, 113, 116, 117, 118, 119, 120, 123, 124, 125, 126, 127, 128, 129, 133, 136, 141, 142, 143, 145, 146, 147, 148, 149, 150, 151, 152, 153, 154, 155, 156, 157
extrusion, 12, 18

F

failure, 172, 174
failure to thrive, 172
families, 103, 104
family, xi, 161, 163, 167, 169
family life, xi, 161, 169
family members, 163
FAO, 170, 179
fat, 10, 11, 14, 53, 55, 56, 85, 171, 172, 173
fatigue, 162
fats, 171
fatty acids, 53
FDA, 76, 77, 78, 79, 86
fermentation, 4, 9, 10, 11, 12, 16, 17, 20, 23, 33, 36, 39, 40, 50, 51
ferritin, 172
fertilization, vii, viii, 27, 28, 29, 30, 31, 32, 33, 34, 35, 36, 37, 38, 39, 40, 42, 43, 46, 47
fertilizers, 29, 40
fiber, xi, 3, 9, 15, 16, 51, 52, 59, 67, 72, 161, 162, 166, 167, 168, 171, 178
fiber content, 3, 59, 72
fibers, 3, 16, 17, 51
fibroblasts, 106
film, 134, 150, 153, 159
film formation, 134, 153
films, 72, 129, 151, 158, 159
filtration, 94, 114, 135
financial, xi, 161
FIQ, 133
flatulence, 168
flavor, 1, 2, 3, 9, 12, 17
flavour, 50, 57, 64, 67
flaws, 50, 81, 82
flexibility, 93, 98, 114, 121, 134, 147, 152, 153, 158
flight, 85
flocculation, x, 32, 113, 115, 117, 118, 120, 123, 124, 125, 127, 129, 130
flow field, 99
fluorescence, 54, 98
foaming, x, 133, 134, 135, 145, 146, 147, 149, 151, 153, 154, 155, 156, 157, 158
foams, 131, 134, 149, 150, 151, 152, 153, 154, 155, 157, 158, 159
focusing, 181
folate, xi, 162, 164, 172, 173, 180
follicle-stimulating hormone, 172
food, viii, ix, x, xi, 2, 7, 15, 19, 47, 70, 71, 75, 77, 78, 79, 80, 81, 82, 83, 84, 86, 87, 89, 102, 106, 113, 114, 132, 133, 134, 157, 158, 159, 161, 164, 165, 167, 169, 170, 173, 175
food additives, 15
food industry, 79, 82, 89, 114
food intake, 173
food production, 47
food products, viii, 7, 19, 75, 77, 80, 102, 169
formation, 4, 5, 12, 17, 21, 29, 33, 35, 36, 37, 38, 62, 66, 67, 68, 71, 72, 90, 92, 93, 94, 96, 103, 106, 108, 115, 120, 126, 131, 134, 138, 139, 145, 150, 153, 157, 174
formula, 12, 14, 22
fracture, 174
fragments, 80, 104
France, 53, 79, 86
free energy, 114
functional food, 25, 70

G

gastrointestinal, 162, 166, 168, 169, 170, 171, 174, 178, 179, 180
gastrointestinal bleeding, 171, 179
gastrointestinal tract, 163, 166
gel, 32, 37, 38, 39, 44, 54, 63, 64, 81, 93, 96, 99
gelatinization temperature, 9
gelation, 114
gender, 169, 173
gene, 176, 181
gene expression, 37, 41
gene therapy, 181
genes, 31, 40, 47, 104, 181
genetic background, 28, 29, 102
genetic factors, 104
genetic linkage, 109
genetic predisposition, 109, 112
genetics, 111
genome, 36, 111
gliadines, vii, 27
global demand, 2
glucose, 3, 10, 13, 14, 21, 24, 70, 172, 175
glucose oxidase, 3, 10, 14, 21, 24, 70
glutamate, 30, 35, 44
glutamic acid, 95, 104, 108, 129
glutamine, 30, 31, 34, 44, 76, 80, 102, 103, 108, 111, 167
glutathione, viii, 27, 34, 36, 38, 39, 43, 44, 45, 46, 96
gluten proteins, vii, viii, ix, x, 5, 18, 27, 28, 29, 30, 31, 32, 33, 36, 37, 38, 40, 45, 49, 50, 76, 81, 82, 84, 87, 89, 93, 94, 96, 97, 98, 99, 100, 101, 107, 108, 114
gluten quantitation, viii, 75, 76, 82
gluten-free bakery, viii, 49, 67
gluten-free bread, vii, viii, 1, 2, 8, 9, 10, 13, 14, 15, 16, 17, 18, 19, 22, 23, 24, 25, 49, 50, 51, 52, 56, 63, 64, 65, 66, 68, 69, 71, 72
gluten-free dough, viii, 9, 15, 16, 22, 25, 49, 63, 72

glutenine, vii, 27
glycemia, 175
glycemic index, 175, 181
grain, 164, 166, 173, 178
grains, 164, 165, 167, 170, 178
granules, 63, 66, 175
groups, x, 133, 137, 142, 143, 145, 153, 156, 157, 164, 169, 177
growth, viii, 28, 29, 30, 31, 34, 38, 40, 43, 44, 46, 87, 110, 162, 172, 174, 175
growth factor, 110
growth rate, 30
guidelines, 70, 174, 175

H

haplotypes, ix, 6, 101, 104
harbors, 108
hardness, 3, 14, 15, 57, 64, 66, 68
harvesting, xi, 77, 161, 168
healing, 168, 172
health, 3, 6, 16, 20, 24, 50, 51, 76, 77, 78, 163, 169, 170, 174, 179
health effects, 76, 77
heart, 166
heart disease, 166
helical conformation, 104
hematological, 172
hemoglobin, 173
hemolytic anemia, 172
hemorrhage, 172
hepatic failure, 106
heterogeneity, 82, 102
high density lipoprotein, 52
high molecular weight (HMW), viii, 27
high-density lipoprotein, 173
high-fat, 173
histamine, 108
histidine, 35
histological, 162, 175
histology, 6, 87, 171
history, 112, 163, 173
HLA, ix, 6, 76, 101, 104, 106, 108, 109, 111, 175
HLA-linked disorders, ix, 101
homeostasis, 42, 109, 176, 181
homocysteine, xi, 162, 173, 180
Homocysteine, 180
homopolymers, 69
hormone, 172, 174
hormone levels, 174
hospitalized, 177
host, 102, 104, 111
HPLC, 134, 136, 139, 141
human, ix, 6, 28, 35, 51, 101, 102, 103, 108, 181

human health, 51
human leukocyte antigen, ix, 6, 101
hybrid, 7, 11, 77, 78
hydrocolloids, viii, 15, 49, 51, 67, 70
hydrogen, ix, 5, 12, 13, 66, 89, 90, 91, 92, 93
hydrogen bonds, 12, 91, 93
hydrogen peroxide, 14
hydrolysates, 135, 136, 139, 141, 142, 143, 150, 151, 155, 157, 158, 159
hydrolysis, ix, x, 21, 113, 114, 115, 116, 119, 120, 124, 129, 130, 131, 132, 133, 134, 135, 136, 137, 138, 141, 143, 145, 147, 150, 152, 153, 155, 156, 157, 158, 159, 175
Hydrolyzates, ix, 113, 116, 119
hydrophilic, 145, 147, 151, 155, 157
hydrophilicity, 145
hydrophobic, 140, 141, 145, 146, 147, 148, 149, 150, 151, 152, 155, 156, 157
hydrophobic interactions, ix, 89, 90, 92, 93, 95, 97, 129, 152
hydrophobic properties, 70
hydrophobicity, 90, 95, 97, 114, 121, 128, 132, 134, 136, 141, 144, 146, 155, 159
hydroxyl, 15
hypercholesterolemia, 173
hyperplasia, ix, 6, 101, 102, 107, 162
hyperprolactinemia, 163
hypertension, 173
hypogonadism, 172
hypothesis, 5, 7, 20, 96, 175
hypothesis test, 96

I

ICAM, 109
ICC, 115, 135
identification, 85, 109, 111, 167, 175
IFN, 106, 176
IFNγ, 176
IL-1, 176, 181
IL-10, 176, 181
IL-15, 176
IL-6, 174
ileum, 172
images, 38, 62
imbalances, xi, 162, 171, 174
immune function, 172
immune response, ix, 76, 77, 101, 102, 107, 108, 109, 172, 176
immunity, 105, 109, 110, 180
immunization, 176
immunogenicity, 104, 105
immunoglobulin, 6
immunological, 162, 168, 176

immunomodulation, 111
immunoreactivity, 86
immunostimulatory, 104, 108, 111
immunosuppression, 167
immunosuppressive agent, 110
immunotherapy, 7, 111
impaired immune function, 172
implementation, 174
impotence, 163
improvements, 15
in vitro, 18, 167, 168, 181
incidence, viii, 49, 172, 175, 176, 177, 178, 180
inclusion, 167, 170
India, 87, 130, 132
indicators, x, 133
indirect measure, 116, 136
individuals, ix, xi, 6, 8, 16, 75, 78, 101, 102, 104, 105, 111, 112, 161, 162, 165, 169, 170, 175, 177
induction, 107, 111, 176
industrialized countries, 76, 77
industries, 164
industry, viii, 3, 28, 51, 75, 76, 78, 79, 82, 84, 85, 89, 114, 168
inert, 164
infancy, 105
infants, 105, 180
infection, 105
infertile, 180
infertility, 162, 177
inflammation, 105, 108, 109, 168, 171, 174, 176
inflammatory, 162, 167, 177
inflammatory bowel disease, 177
inflammatory cells, 107
inflammatory disease, 75, 102, 105
inflammatory response, 162
infrastructure, 8
ingest, 166
ingestion, ix, 101, 102, 163, 165, 167, 178, 179
ingredients, ix, x, 3, 9, 12, 18, 50, 53, 55, 56, 67, 68, 78, 79, 82, 113, 114, 133, 134, 164
inhibition, 106, 110, 176
inhibitor, 176
initiation, 105, 173
injections, 54
insulin resistance, 175
integration, 52
integrins, 106
interaction, 181
interaction effect, 13
interaction effects, 13
interactions, 153, 175, 180
interface, x, 114, 115, 118, 121, 125, 132, 134, 147, 150, 151

interfacial layer, 119
interfacial tension, 134
interferon, 105, 108, 109, 176
interferon (IFN), 176
interferon-γ, 108
interleukin, 174, 181
Interleukin-1, 176
international trade, viii, 75
internet, 169
intestinal tract, 103, 108
intestine, ix, 6, 17, 75, 101, 104, 108, 111, 163, 171, 173, 174, 175
intraepithelial lymphocytosis, ix, 6, 101
intrauterine growth retardation, 162
inversion, 115, 131
investment, 2
investments, 2, 3
ion uptake, 42
ionization, 81, 85
iron, xi, 15, 22, 52, 72, 85, 161, 162, 164, 166, 171, 172, 180
iron deficiency, 171, 172
irritable bowel syndrome, 167
isolation, 99
isoleucine, 95
isotherms, 158
issues, 21
Italy, 56, 161

J

jejunum, 172

K

kinetics, 68, 118

L

labeling, viii, 75, 76, 77, 78, 79, 83, 86
lactase, 14, 165
lactase deficiency, 165
lactose, 165, 167
lactose intolerance, 165
lamella, 129
lamellae, 152
lamina, 162
lamina propria, ix, 101, 107, 108, 109, 162
lead, xi, 28, 41, 51, 59, 64, 95, 97, 102, 105, 110, 115, 129, 161, 162, 163, 175
lecithin, 11
legislation, 76, 78
lesions, 102, 103, 107, 172
leucine, 35, 95
leukocyte infiltration, ix, 101

lifestyle, xi, 161, 163, 169, 178
light, 28, 99, 117, 118, 130, 131, 175
light scattering, 99, 130
lipid, 173, 175, 180
lipid metabolism, 173, 175
lipid profile, 180
lipids, xi, 69, 161
lipoprotein, 173
liquid chromatography, 37, 54, 79, 81, 99
liquid interfaces, 131, 158
liquids, 54, 114
lithium, 54
liver, 106
liver disease, 106
LM, 177
loaf volume, vii, 11, 12, 15, 27, 28, 32, 33, 37, 39, 43
localization, 108
locus, 104
London, 157, 158
longitudinal study, 177
loss of libido, 163
low temperatures, 97
low-density, 173
lower lip, 59, 68
LSD, 119, 138
lumen, 107, 175
Luo, 32, 37, 44
luteinizing hormone, 172
lymphocyte, 176, 181
lymphocytes, 106, 108, 109, 162, 168, 181
lymphocytic infiltration, ix, 101
lymphocytosis, ix, 6, 101, 109
lymphoma, 109, 110, 162, 167, 170
lymphomagenesis, 181
lysine, 35, 36, 51, 95, 104

M

machinery, 77
macrocytosis, 172
macromolecules, 107, 115
macronutrients, xi, 162
macrophage, 176
macrophages, 109, 176
magnesium, xi, 52, 162, 172
Maillard reaction, 66, 68, 72
major histocompatibility complex, 104
majority, 79, 103, 109, 168
malabsorption, 163, 167, 171, 172, 180
malignancy, 103, 162
malnutrition, 167
management, vii, 27, 28, 40, 70, 165, 167, 170, 173, 177, 179
manufacturer, 164

manufacturing, 2, 3, 5, 8, 9, 11
margarine, 164
market, x, 133, 134
mass, 4, 5, 56, 67, 79, 81, 83, 85, 103, 173, 174, 175, 181
mass spectrometry, 79, 81, 83, 85
materials, 50, 52, 55, 57, 59, 67, 71
matrix, 13, 16, 37, 40, 50, 54, 63, 82, 85, 103, 109
matrix metalloproteinase, 109
matter, iv, 42, 77, 95
meals, 163, 164
measurement, 53, 56, 80, 86, 116, 117, 136, 165, 174
measurements, 25, 40, 52, 53, 54, 55, 56, 57, 59, 63, 65, 67, 68, 130, 138, 174
measures, 174
meat, 77, 173
mechanical properties, 62, 66, 67, 98
media, 93, 115, 119, 120
medical, ix, 101, 165, 177
medications, 164, 171
medicine, 17
Mediterranean, 42
mellitus, 181
membranes, 43, 80
men, 163, 169, 178, 180
menarche, 163
menopause, 162, 163
metabolic, xi, 162, 173
metabolism, 30, 34, 36, 37, 45, 172, 173, 174, 175, 181
metabolites, 29, 34, 40
metals, 34
methionine, vii, 27, 34, 35, 36, 40, 43, 45, 51
methodology, 11, 24, 25, 71, 72, 73, 81, 130, 158
methyl cellulose, 15, 16
methylcellulose, 72
microarray, 181
microscopy, 99
microstructure, 18
Middle East, 75, 102
migration, x, 106, 113, 115, 118, 176
mixing, 4, 5, 9, 11, 12, 20, 24, 32, 40, 53, 54, 56, 57, 77, 115
Mixolab, viii, 49, 52, 53, 55, 57, 58, 59, 63, 67, 71
ML, 179
model system, 68
models, 36, 111
modifications, iv, vii, 1, 3, 12, 82, 84, 98, 118, 133, 157, 161, 175
modulation, 176
modulus, 56, 63, 65, 66, 90
moisture, 3, 11, 16, 42, 53, 56, 95, 96, 98, 100, 115, 135

Index

moisture content, 3, 16, 53, 95, 100
mold, 9
molecular mass, 81
molecular structure, 52, 158
molecular weight, viii, 4, 27, 29, 37, 61, 80, 91, 92, 93, 94, 96, 97, 100, 103, 114, 121, 123, 127, 136, 140, 141, 142, 146, 147, 148, 149, 151, 152, 155, 159
molecular weight distribution, 91, 97, 159
molecules, viii, 27, 29, 36, 65, 66, 76, 90, 92, 103, 106, 109, 134, 141, 143, 145, 154, 156
monoclonal antibody, 86
monozygotic twins, 104
morbidity, xi, 102, 161, 162, 163
morphology, 6, 54, 114, 171
morphometric, 178
mortality, xi, 7, 102, 161, 162, 167, 173, 177, 178
mortality rate, 7
mucosa, 6, 13, 102, 103, 106, 107, 108, 109, 110, 162, 168, 171
multiple myeloma, 86
muscle, 163, 172
muscle weakness, 163, 172
musculoskeletal, 172
musculoskeletal pain, 172

N

Na^+, 70
NaCl, 29, 55, 76
naphthalene, 95
native starches, 11
natural, 181
natural killer, 181
natural killer cell, 181
necrosis, 109, 174
negative effects, 35, 78, 102
Netherlands, 43, 45, 70, 171, 179, 180
network, 175
neural network, 85
neuropathy, 172
neutral, 67, 85, 114, 127, 134, 151
neutralization, 176
neutropenia, 172
nitrogen, vii, x, 27, 28, 41, 42, 43, 44, 46, 47, 103, 133
Nitrogen fertilization, vii, 27, 31, 36
non-Hodgkin lymphoma, 162
non-Hodgkin's lymphoma, 162
non-immunological methods, viii, 75
non-polar, 93
normal, 168, 169, 170, 173, 175
normalization, 168
North America, 166, 171, 178

nucleic acid, 172
nutrient, 28, 31, 40, 43, 51, 163
nutrients, viii, 30, 31, 40, 46, 49, 51, 103, 166
nutrition, 28, 35, 41, 42, 43, 46, 47, 165, 167, 173, 177, 180
nutritional imbalance, xi, 161, 171
nutritional screening, 172

O

oil, x, 12, 15, 16, 56, 113, 114, 115, 116, 117, 119, 120, 121, 126, 130, 131, 132
oligomers, 44
oligospermia, 172
olive oil, 15
opportunities, 175
optimization, 15, 72
oral, 175
oral health, 24
organ, 34
organs, viii, 27, 30
oscillation, 63
osteomalacia, 163, 172, 174
osteopenia, 163, 174
osteoporosis, 163, 170, 173, 174, 177
overlay, 38
overweight, xi, 162, 171, 173
oxidation, 10, 13, 94
oxidative stress, 34
oxygen, 13, 44

P

PA, 179
pain, 163, 169, 170, 172, 174, 180
pancreatic, 167
pancreatic insufficiency, 167
parallel, 56
pasta, 2, 5, 18, 164, 171
pathogenesis, ix, 21, 75, 86, 87, 101, 102, 104, 106, 107, 108, 110, 112, 176, 181
pathogens, 28
pathophysiological, 102
pathophysiology, 163
patients, xi, 161, 162, 163, 164, 165, 166, 167, 168, 169, 170, 171, 172, 173, 174, 175, 176, 177, 178, 179, 180, 181
PCR, 79, 81, 82, 83, 87
pediatrician, 8
PEP, 17
peptide, x, 76, 80, 89, 104, 107, 108, 110, 114, 115, 119, 120, 123, 129, 133, 135, 137, 139, 141, 142, 145, 147, 151, 157
Peptide, 136, 137, 142

peptide chain, x, 133, 142, 157
peptides, x, 76, 77, 79, 80, 97, 102, 103, 107, 108, 109, 110, 112, 113, 114, 121, 123, 126, 127, 129, 130, 131, 132, 136, 139, 141, 143, 146, 147, 149, 151, 153, 155, 156, 157, 159, 176, 181
perceptions, 169
peripheral blood, 104, 111
permeability, 7, 105, 107, 111, 176
pH, ix, x, 30, 54, 68, 103, 113, 114, 116, 117, 118, 119, 120, 122, 123, 124, 125, 126, 127, 128, 129, 132, 133, 134, 135, 136, 137, 139, 140, 141, 142, 143, 144, 145, 146, 147, 148, 149, 150, 151, 152, 153, 154, 155, 156, 157
pharmaceutical, 178
phase boundaries, 121
phase inversion, 131
phenolic compounds, 16
phenylalanine, 80, 95
phloem, 30, 34, 36, 41
phosphate, 32, 137
phosphorous, 172
phosphorus, 20
phosphorylation, 108
physical health, 169
physical properties, 5, 63, 99
physicians, 170
physicochemical, 134, 139
physicochemical properties, 70, 97, 100, 134
physics, 99
Physiological, 41
placebo, 85, 179
plant growth, viii, 28, 29
plants, vii, 19, 27, 28, 30, 31, 33, 35, 36, 37, 38, 39, 41, 43, 45
plasma, xi, 162, 173
plasma cells, 106, 109
plasmids, 176
play, 175, 176
polar, 81, 91, 93, 95, 134
polar groups, 91, 93
polarity, 81
polyacrylamide, 80, 99
polydispersity, 117
polymer, 36, 39, 52, 91, 92, 93, 94, 96, 97
polymerase, 81, 87
polymerase chain reaction, 81, 87
polymerization, 38
polymers, 39, 44, 91, 95, 96, 97, 100
polypeptide, 46
polypeptides, 34, 93, 126, 141, 143, 157
polyphenols, 25, 52
polysaccharide, vii, 1, 64
poor, 143, 169, 171, 172, 173, 180

population, viii, xi, 2, 75, 76, 104, 109, 119, 120, 161, 162, 175, 177, 179, 180
porosity, 14
positive relationship, 39
potassium, 94
potato, 11, 13, 16, 72
potatoes, 171, 173
potential benefits, 176
precipitation, 31
pre-eclampsia, 172
pregnancy, 177, 180
preparation, iv, 10, 28, 32, 54, 67, 82, 83, 127, 131, 158, 165
private enterprises, vii, 1
probe, 56, 95
production, 173, 181
professionals, 78
program, 179
project, 45, 84
proliferation, 103
proline, 76, 80, 95, 102, 103, 111, 167
prolyl endopeptidase, 103, 111, 175, 176
promoter, 109
proteases, x, 133
protection, 103
protective factors, 105
protein, x, 133, 134, 135, 136, 137, 138, 139, 141, 142, 143, 145, 147, 148, 149, 150, 151, 152, 153, 155, 157, 158, 159, 170, 172, 174, 175, 181
protein hydrolysates, 135, 159
protein network structure, vii, 1, 67
protein structure, 24, 29
protein synthesis, 29, 31, 34, 37, 38, 45, 172
protein-protein interactions, 134, 151
proteolysis, 104, 145, 159
proteolytic enzyme, 24, 115, 135
proteome, 47
proteomics, 42
prothrombin, 172
pseudo-cereal-based formulations, viii, 49
puerperium, 177
purification, 19
purity, xi, 161
pyridoxine, 52

Q

quality of life, xi, 7, 10, 161, 169, 170, 178, 179, 180
Quality of life, 179
quantification, 82, 83, 132
quercetin, 52
questionnaire, 169, 170

R

range, x, 133, 135, 136, 173
raw materials, 50, 52, 71
RDA, 171
reaction zone, 92
reactions, viii, 6, 10, 13, 31, 43, 49, 66, 68, 72, 90, 94, 99, 100, 106, 143
reactive groups, 90, 114
reactivity, 23, 85, 111
reading, 117, 164
recall, 163
receptors, 106, 109, 110
recognition, 80, 102, 109, 110, 112
recommendations, iv, 166, 167, 178
recovery, 13, 15, 23, 31, 79, 85, 165, 175
red blood cell macrocytosis, 172
redistribution, 46, 110
reduction, 143, 144
refractory, 163, 165, 167, 171, 175
regenerate, 30
regions of the world, 162
regulation, 168, 174, 175, 176
regulations, viii, 44, 75, 79
reinforcement, 163
relapse, 168
relapses, 8, 79
relationship, 173
relationships, 152, 158, 159
relatives, 104, 177
reliability, 83
reparation, 28
reproduction, 172, 180
repulsion, 126, 129
requirements, 35, 180
researchers, 2, 3, 10, 52, 90, 91, 92, 94, 105, 110
residues, 14, 38, 40, 50, 76, 93, 94, 95, 96, 103, 108, 134, 141, 143, 145, 167
resilience, 13
resistance, 3, 9, 28, 32, 33, 38, 39, 40, 94, 96, 104, 128, 134, 175
response, ix, 10, 11, 24, 25, 44, 46, 71, 72, 76, 77, 101, 102, 107, 108, 109, 110, 111, 130, 158, 162, 165, 168, 172, 173, 175, 176, 181
restaurant, 164
restaurants, 164
retardation, 153, 162, 172
retention, 141
rheological property, viii, 28
rheology, 11, 18, 20, 52, 70, 91, 95, 97, 98, 100, 130
rheumatoid arthritis, 162
riboflavin, xi, 52, 72, 162, 174
rice, 164, 171
risk, xi, 7, 70, 76, 77, 82, 104, 105, 109, 110, 112, 162, 163, 168, 169, 173, 174, 175, 176, 177, 179, 180
risk factors, 70, 163, 175
risks, 162
room temperature, 11, 56, 57, 116, 136
root, 9, 30, 33
roots, 8, 9, 30, 31, 33, 34, 43, 44, 45, 47, 52
rotavirus, 105
routes, 108
Royal Society, 98, 99, 100
RP-HPLC, x, 133, 136, 139, 140, 141, 144, 146, 147, 148, 149, 151, 152, 153
rye, x, 161, 162, 167

S

safety, viii, xi, 16, 75, 77, 78, 83, 84, 110, 161, 164, 167, 168, 171
salt, 159
sample, x, 133, 135, 136, 137, 138, 147
saturated fat, 171
saturation, 119
scanning electron microscopy, 52, 54, 67
scattering, 99, 117, 123, 130
school, 164
science, 52
scientific progress, 2, 7
scientific publications, 102
secretion, 109, 172
sedimentation, 32, 39, 115, 117, 118
seed, 15, 18, 41, 45, 46
seedlings, 28
seeds, 173
selenium, 172, 180
self-care, 169
semen, 180
sensitivity, x, 6, 21, 81, 102, 161, 162, 171
sensitization, 111, 162
serological marker, 167
serology, 6
serum, 6, 51, 85, 106, 137, 158, 163, 172, 174
serum albumin, 158
serum ferritin, 172
shelf life, 10, 51, 69
short-term, 168
showing, 10, 17, 63, 65, 92, 154
sign, 152, 165, 171
signalling, 107
signals, 83
significance level, 57
signs, 6, 106, 168
Singapore, 98
sites, 167

skeleton, 175
skin, 172
small intestine, ix, 6, 75, 101, 108, 163, 171, 174, 175
smoking, 163
social activities, 169
social problems, 111
sodium, 51, 137
software, 54
solubility, ix, x, 4, 62, 77, 82, 90, 91, 93, 96, 97, 98, 113, 114, 133, 136, 138, 139, 141, 143, 157
solution, x, 4, 29, 80, 82, 84, 96, 114, 116, 133, 134, 136, 137, 138
solvents, 90, 97
Sorghum, 165
South Africa, 7, 19, 51
sowing, 35, 37, 38, 42
species, 13, 30, 50, 78, 82, 87
specific surface, 117
spectrum, 177
spontaneous abortion, 172
sprouting, 18
sprue, ix, 101, 102, 109, 163, 167, 177, 181
S-rich gluten proteins, vii, 27, 34, 36, 39
SS bonding, ix, 89, 90, 92, 97
SSA, 25
stability, x, 16, 58, 67, 70, 91, 98, 99, 113, 114, 115, 117, 119, 121, 123, 129, 130, 131, 132, 134, 135, 138, 149, 150, 151, 152, 153, 155, 156, 157, 158, 159
stabilization, 42, 131, 151, 165
stabilizers, 164
standard deviation, 63, 64, 66, 117, 120
standardization, 84
Standards, 179
starch, vii, 1, 5, 9, 10, 11, 12, 13, 14, 15, 16, 17, 19, 21, 22, 23, 37, 40, 50, 53, 57, 59, 63, 66, 67, 68, 69, 71, 78, 82, 90, 91, 96, 98, 103, 114, 115, 134, 135, 164, 175, 179, 181
starch granules, 63, 66, 175
starvation, 37
state, 4, 90, 92, 95, 163, 172
states, 95
statistical analysis, 138
statistics, 6
steatorrhea, 173
stomach, 171
storage, viii, 8, 9, 14, 15, 16, 17, 22, 28, 29, 30, 31, 33, 41, 45, 46, 50, 51, 56, 57, 63, 66, 77, 98, 103
strategies, 166, 175
strength, 134, 151
stress, 34, 106, 109
stretching, 52

strong interaction, 114
structural changes, 82
structural characteristics, 115, 121, 135
structural modifications, 84
structural protein, 29, 30
structural transformations, 114, 134
structure, vii, ix, 1, 12, 16, 20, 21, 29, 39, 50, 52, 62, 63, 64, 65, 67, 72, 82, 89, 90, 91, 92, 93, 94, 96, 99, 129, 134, 149, 153, 158
structure formation, 67
structuring, 51
substitutes, 2, 11, 17
substitution, 3, 8, 50
substrate, 80, 104, 106, 114, 115, 135, 137
substrates, 7, 104, 143
sulfate, viii, 27, 33, 34, 41, 42, 45
sulfur, vii, 27, 28, 41, 43, 44, 45, 46, 47, 91
sulphate, 171
sulphur, 41, 42, 43, 44, 46, 47, 92
Sun, 16, 25
sunflower, 173
supplementation, xi, 162, 164, 171, 172, 174, 180
supplements, 163, 164, 166, 171, 174, 180
supply, 173
suppression, x, 113, 152
surface area, 117, 121
surface properties, 114, 119, 134, 138
surface tension, 151
surfactant, 131
surfactants, 11, 12, 114, 134
susceptibility, ix, 6, 101, 105
suspensions, 52
Sweden, 86
sweets, 173
swelling, 10, 11, 40, 59, 66, 68, 69
switching, 171
symptom, 179
symptoms, xi, 6, 7, 31, 37, 40, 161, 162, 163, 165, 166, 168, 169, 170, 172, 178, 180
syndrome, 167
synthesis, viii, 27, 29, 30, 31, 34, 36, 37, 38, 41, 42, 45, 46, 172, 173, 181

T

T cell, 104, 106, 107, 108, 109, 110, 111, 162, 168, 176, 181
T cells, 162, 176, 181
T lymphocyte, 181
T lymphocytes, 109, 181
target, 80, 106, 108, 176
taxonomy, 103
T-cell, 162, 168, 175, 176
TCR, 107

Index

technical assistance, 41
techniques, 38, 57, 76, 80, 81, 90, 98, 111
technology, 2, 22, 71, 111, 176
temperature, 8, 9, 11, 12, 53, 56, 57, 68, 90, 91, 92, 93, 94, 96, 97, 98, 99, 114, 115, 116, 121, 135, 136
tension, 115, 121, 134, 151
testing, 14, 68, 71, 79, 85, 111, 179, 180
texture, 2, 3, 9, 10, 11, 12, 14, 16, 17, 50, 72, 99, 134
TGF, 106
therapy, ix, xi, 102, 162, 175, 181
thermal stability, 91
thermal treatment, 82, 115, 135
thinning, 52
three-dimensional polysaccharide, vii, 1
threonine, 35
threshold, xi, 161, 171
thymus, 106
time, 138, 139, 141, 150, 151, 152, 154, 155, 156
tissue, 6, 29, 30, 41, 76, 102, 105, 107, 108, 109, 110, 162, 165, 167, 172, 174, 178
TNF, 174
total cholesterol, 52, 173
total plasma, xi, 162, 173
toxic, 164, 181
toxic effect, 111
toxicity, 7, 13, 19, 80, 82, 84, 86, 103, 167, 168, 170, 178, 179
trade, viii, 75
transferrin, 108
transformations, 114, 134
transforming growth factor, 110
transglutaminase, 157, 165, 167, 174, 178
translocation, 23, 30, 31, 34, 36, 38, 45, 46
trans-membrane, 172
transmission, 117
transport, viii, 27, 34, 41, 42, 43, 45, 77, 108, 172, 181
transportation, 8, 77, 172
travel, 169
treatment, ix, xi, 2, 3, 7, 21, 23, 40, 76, 85, 89, 90, 92, 93, 96, 99, 101, 110, 115, 130, 135, 157, 161, 162, 163, 165, 166, 169, 170, 171, 173, 175, 177, 178, 179, 181
trial, 21, 85, 87, 178, 179
trichloroacetic acid, 135, 136
trifluoroacetic acid, 136
triggers, 17
triglyceride, 175
troubleshooting, 165
tryptophan, 95
tumor, 109, 174
tumor necrosis factor, 109, 174

type 1 diabetes, 175
type 2 diabetes, 178

U

UK, 21, 71, 117
ultrasound, 130
ultrastructure, 22
un-hydrolyzed, x, 133
uniform, 64, 115, 135
United, 6, 7, 19, 20, 24, 45, 85, 86, 162, 170, 177
United Nations, 170
United States, 6, 7, 19, 20, 24, 45, 85, 86, 162, 177
universities, 3
urea, 137
US Department of Health and Human Services, 86
USA, xi, 43, 51, 54, 57, 68, 71, 72, 100, 117, 132, 161, 177

V

vaccine, 176
Valencia, 131
valine, 35, 95
valuation, x, 113
values, 143, 145, 147, 152, 175
valve, 119
variability, 170
variables, 102, 115
variations, 9
varieties, 45, 47, 50, 76, 78, 91, 94, 96, 97, 98, 111
vascular disease, 180
vegetable oil, 12, 16
vegetables, 164, 173
vertical scan macroscopic analyzer, x, 113
viscoelastic properties, ix, 5, 9, 12, 20, 50, 51, 56, 65, 98, 113, 128
visco-elastic properties, viii
visco-elastic properties, 49
viscosity, 8, 11, 12, 52, 56, 59, 67, 68, 103, 114, 115, 121, 127, 151
vitamin A, 172
vitamin B1, 172, 174
vitamin B12, 172, 174
vitamin B12 deficiency, 172
vitamin B6, 173
vitamin C, 16
vitamin D, 163, 164, 166, 172, 174
Vitamin D, 174, 180
vitamin D deficiency, 174
vitamin E, 172, 180
vitamins, xi, 51, 162, 164, 166, 172

W

Washington, 131
water, 3, 4, 5, 9, 10, 11, 12, 16, 24, 29, 53, 54, 55, 56, 58, 67, 70, 76, 84, 90, 93, 103, 114, 115, 116, 121, 130, 131, 132, 134, 135, 136, 150, 151
water absorption, 55, 58, 67, 103
weakness, 163, 172
weight gain, 163, 173
weight loss, 167, 173
weight management, 173
western blot, 80, 83
wetting, 5
wheat, x, 133, 134, 150, 151, 155, 157, 158, 159, 161, 162, 165, 167, 168, 175, 176, 178, 179, 181
wheat bread, vii, 1, 9, 10, 11, 12, 13, 14, 15, 17, 42, 70
wheat plant development, vii, 27, 36
WHO, 3, 170, 173, 179
whole grain, 178
withdrawal, 109, 168
women, 163, 169
workers, 77, 91
World Health Organization, 2, 170
World Health Organization (WHO), 3, 170
worldwide, xi, 6, 161, 162
wound healing, 172

X

xanthan gum, 13, 15, 16, 51, 70
xylem, 30, 34

Y

yeast, 9, 16, 22, 25, 53, 55, 56, 73, 176
yield, 3, 8, 9, 28, 30, 31, 40, 42, 43, 44, 47
yogurt, 166
yolk, 130
young adults, 180
young people, 70, 179

Z

zinc, xi, 162, 172, 180